NICK MIDDLETON

Nick Middleton is a geographer, writer and presenter of television documentaries. He teaches at Oxford University, where he is a Fellow of St Anne's College. A Royal Geographical Society award-winning author, he works, teaches and communicates on a wide variety of geographical, travel and environmental issues for a broad range of audiences, from policy makers to primary-school children. He is the author of seven travel books, including the bestseller GOING TO EXTREMES, *which accompanied a television series he wrote and presented for Channel 4 on extreme environments and the people who live in them.*

A Compendium of Fifty Unrecognized and Largely Unnoticed States

NICK MIDDLETON

MACMILLAN

First published 2015 by Macmillan
an imprint of Pan Macmillan
20 New Wharf Road, London N1 9RR
Associated companies throughout the world
www.panmacmillan.com

ISBN 978-1-4472-9527-3

9 8 7 6 5 4 3 2 1

A CIP catalogue record for this book is available from the British Library.

Designed and typeset by Sarah Greeno www.sarahgreeno.com
Map artwork by Sarah Greeno
Printed and bound in Malaysia

This book is for John Ellis Middleton
(1924–2013)

CONTENTS

Map symbols 9

Introduction 11

Europe 18
Africa 72
North America 106
South America 132
Asia 142
Oceania 192
Elsewhere 214

Acknowledgements 232

MAP SYMBOLS

The maps within this Atlas make use of the following symbols:

Existing international boundary

International boundary of unrecognized country

State boundary

Capital city

Major settlement

River

Body of water (sea, major lake)

Road

Railway

Mountain

Volcano

A note on flags

Each would-be country is presented with its national flag. In a few cases, two flags are presented, each representing a different separatist group.

INTRODUCTION

Leopold II, King of the Belgians, was known for his prodigious appetite. He frequently ordered another entrée after finishing an enormous meal, and once ate two entire roast pheasants at a Paris restaurant. It is not surprising, therefore, that he used a culinary metaphor when declaring his determination during the nineteenth-century scramble for African territory to obtain the largest possible slice of what he called the 'magnificent African cake'.

At the Berlin Conference on Africa in 1885, Leopold secured his own private colony seventy-five times larger than Belgium, as Europe's leading powers carefully divided up the entire continent between themselves. Studying a 5-metre-high wall-map of Africa, the diplomats agreed the ground rules for taking possession of its territory, and began negotiating the boundaries between their various colonies. And so concluded the final phase of the global process of European colonization that had begun more than 300 years earlier with Spanish and Portuguese explorers.

Everybody knows what today's political map of the world looks like. The bold colours and sharp boundaries show the global land surface neatly divided between sovereign states. But it hasn't always been like this. For most of human history, before the Europeans started exploring and colonizing, people lived in small cultural communities or larger civilizations that were hardly interlinked at all. With time, as more people moved more frequently and more quickly – exploring, conquering, trading and travelling – so the contemporary world of countries, tightly defined by their boundaries, developed.

The final phase of this process is really quite recent. It is only after the end of World War II, with the creation of the United Nations and the process of decolonization, that we came anywhere near to the map of many colours we know today. A truly global international society of countries.

Not that the political world map is static. Countries come and go. Towards the end of the twentieth century, the disintegration of the Soviet Union spawned no fewer than fifteen new states and East Germany joined its western counterpart to become a reunified country. These were quickly followed by Czechoslovakia undergoing a

'Velvet Divorce' to create the Czech Republic and Slovakia. Already in the twenty-first century we have seen more new states emerge in Asia (East Timor), Europe (Montenegro) and Africa (South Sudan).

But at the same time, we are constantly being reminded that we live in an era of unprecedented global communication, a time when globalization is eroding the importance of the nation state. Our planet is becoming an increasingly borderless place, where national boundaries matter little to the movement of goods and investment (though the movement of migrants is another story). National governments have had their power diluted and usurped by some new actors on the global stage, including international organizations, transnational corporations and non-governmental organizations, or NGOs. A world of fixed spaces is giving way to a world of flows, and the idea of national territory is giving way to supra-national communities such as the European Union. With its echoes of Aldous Huxley, this is the 'New World Order'.

However, while the notion of fixed territories is in one sense under threat from globalization, the rise of the internet, virtual communities and the diffusion of ideas, there is no question that the national space itself remains of great importance. Individual countries still dominate all of our lives. Much as some might like to think of themselves as 'Citizens of the World' rather than citizens of any one nation state, they won't get very far in seeing that world without a travel document issued by their national government. Granted, the European Union has, to a large extent, done away with its internal boundaries, but the EU is still a relatively small chunk of the world. An EU citizen who ventures outside the EU can only do so legally with a passport.

Which brings us back to that political map of the world. Announcements of the end of the nation state may be premature. National territory still has an enduring allure. And nation states work hard to keep it that way, defending borders and encouraging many schemes to strengthen national cohesion.

What most of us probably don't realize about that world map is what it conceals: a multitude of unrecognized and largely unnoticed states whose claims to legitimacy are made invisible by the bold, self-assured slabs of colour. This is the shadowy, surprisingly large, and literally unofficial world of countries that don't exist.

This Atlas presents fifty of these wannabe nation states. Each has its own flag and legitimate claim to some territory but, for a variety of reasons, none has quite made the grade, to join the exclusive club of internationally recognized countries.

WHAT IS A COUNTRY?

Selecting which non-countries to include in this book was complicated by a lack of consensus on what exactly constitutes a country. The concept is old, but also notoriously slippery. As soon as you set out to find a clear definition you start running into discrepancies, exceptions and anomalies.

An apparently straightforward answer might be that all 'real' countries have a seat at the General Assembly of the United Nations, the world's most important and prestigious state-based international organization. That certainly covers most generally accepted countries of the world, but it is not a definitive solution. Israel became a member of the world body in 1949, but more than thirty other UN member states, from Cuba and Bangladesh to Morocco and Saudi Arabia, do not recognize Israel's existence.

To complicate matters further, the UN recognizes other countries that do not have full membership. In 2012, Palestine joined the Holy See to become a non-Member Observer State at the UN. However, when the UN recognized the state of Palestine, not all of its members agreed. Some still refuse to recognize Palestine as a country at all. Interestingly, the Holy See is not actually a country either. The Holy See is, in effect, the pope, or at least his office — the papacy — and not the Vatican, the small state where it is based.

Even full UN membership is not necessarily a guarantee of country status. A case in point is TAIWAN *(here in capitals, indicating that the state appears in this Atlas as one of the 'countries that don't exist'). In the early years of the United Nations, most countries recognized the Chinese regime in Taiwan (also known then as 'Free China') while the mainland communists (or 'Red China') were isolated diplomatically. Abruptly, in 1971, this absurd situation was reversed. Since then, Taiwan has been forced to operate in the diplomatic twilight, running numerous 'trade offices' around the world, but very few official embassies.*

And then there are the countries that only have UN membership when considered as a group. The type example here is the United Kingdom, or the United Kingdom of Great Britain and Northern Ireland, to give it its full title. This is the collective name of four countries: England, Wales, Scotland and Northern Ireland. These four separate countries are united under a single parliament through a series of legal 'Acts of Union'. So when it comes to international relations, England, Wales, Scotland and Northern Ireland are all represented by the UK government.

In many international sporting events, however, it is a different matter. Through a quirk of history, all four have separate teams for football and rugby and several other sports. Unless the competition is the Olympics, in which case the British Olympic Association fields teams and individuals to represent the United Kingdom.

Still more confusion is introduced by the more-or-less interchangeable use of the terms 'country' and 'state'. Interestingly, the United Nations itself uses neither, but introduces another term — 'nation' — also frequently used to mean the same. Some authorities prefer to reserve the word nation for a social, ethnic or cultural group that might have its own country, also sometimes called a 'nation state'. Hence, Israel is frequently referred to as the Jewish state or nation state, although in reality many non-Jewish people also live there.

A widely accepted legal definition of a state was hammered out at a meeting in Uruguay in the 1930s. Article 1 of the Montevideo Convention sets out the four essential criteria for statehood: a permanent population; a defined territory; a government; and the capacity to enter into relations with other states.

That sounds fine, although we have already seen that 'the capacity to enter into relations with other states' is by no means always enough. One territory, NORTHERN CYPRUS, *has been recognized only by Turkey and no other states at all. For this reason, it has been included in this Atlas.*

Many authorities on the subject would also include the idea of power in their definition of a country. The German social and political scientist Max Weber defined statehood in blunt terms: as a monopoly of the legitimate use of violence over a given territory. Certainly violence has helped many countries gain and hold on to territory, and continues to be a potent symbol of national strength. But countries are also recognized as wielding other forms of power: by making and upholding the norms and rules that apply within their boundaries.

Almost immediately, as usual, anomalies appear. The universally recognized sovereign state of Somalia has singularly failed to maintain control over most of its territory since descending into a chaotic civil war in 1990. Yet for most of that time, the northern part of the country — SOMALILAND *— has managed to maintain law and order within its own borders. Since declaring independence in 1991, Somaliland has simply carried on regardless. It has all the hallmarks of a fully fledged country: its own parliament, currency, car registrations, even biometric passports. Yet unlike its chaotic neighbour, it has not been recognized by any other state.*

WHAT'S IN . . . WHAT'S NOT?

This is not a definitive compendium. This Atlas could have been filled several times over with 'nations in waiting', but a selection of fifty has been made based on some rules of thumb. All of the 'non-countries' included have failed to secure a seat at the United Nations General Assembly and none has widespread international recognition as a sovereign state. All of these 'non-countries' have at least the outward trappings of national consciousness, including a flag, some form of government and a claim to territory, as well as a seriousness of purpose.

Some of the entries have in fact established their exclusive control over territory for considerable lengths of time, despite their lack of international recognition. Taiwan and Somaliland fall into this category: both 'de facto states' that wait only for the rest of the world to come to terms with the reality of their existence.

Many currently exist as partially autonomous regions of larger recognized states. Their claim to greater areas and greater levels of self-determination are frequently based on historical precedent, treaty or an ethnic/cultural distinctiveness that puts them apart from those who dominate the state in which they live. Their likelihood of gaining more territory, or greater autonomy, varies right across a spectrum from very unlikely (e.g. CABINDA, LAKOTAH, TIBET*) to quite possible (e.g.* GREENLAND*).*

Others are territories that have been declared independent by individuals or small groups and have a minimal chance of being recognized as independent by any established nation state or international body. These so-called 'microstates' are usually small in either land area and/or population (e.g. PONTINHA*). Some, such as* FORVIK *and* HUTT RIVER*, are deliberate parodies, designed to mimic the fully fledged state but with a serious personal, political and/or commercial ambition.*

Some would-be nation states included here owe their survival to the support of one major ally. Many splinters of the former Soviet Union fall into this category, including TRANSNISTRIA *and* ABKHAZIA, *which are supported by Russia. Indeed, there are many other fragments of the former Soviet Union that could have been included in this Atlas, but their yearnings for self-rule are essentially similar in nature to those two examples.*

Some of those with few allies have joined forces with other unnoticed states to form their own parallel international community. Transnistria has made a link with Abkhazia and some other forgotten Soviet flashpoints, to create the Community for Democracy and Rights of Nations. A central aim of the organization, also dubbed the Commonwealth of Unrecognized States, appears to be recognition of each other's independence. Other, larger bodies with similar agendas include the Unrepresented Nations and Peoples Organization (UNPO) and the Unrepresented United Nations (UUN).

*To balance the common theme of desire for independence, a few examples have been included of territories that chose to be subsumed by another (*TUVA *decided to join the USSR and remains a part of Russia), or regions that prefer to remain a colony rather than to go it alone as an independent entity (*MAYOTTE*).*

Many islands are represented in these pages, and this is no coincidence. Being physically separate by virtue of being an island makes life as a sovereign entity much easier. Indeed, some microstates have even set themselves up on newly created islands (e.g. MINERVA, SEALAND*).*

All of these unrecognized countries are, to some degree, unique entities. This fact, when combined with the realization already mentioned that there is no one universally acknowledged definition of a country, means that this compendium of non-states is, to some extent, inherently arbitrary. Certain countries that some readers might expect to find in these pages have been excluded for a variety of reasons. Both Israel and Palestine, as mentioned above, do not exist in the eyes of a number of states. But likewise both have many more supporters who do recognize them, and therefore neither state appears in this book.

Any one, or all of the four constituent countries of the United Kingdom could have found their way into this Atlas given their lack of a UN seat. But all have been omitted, simply because each is sufficiently recognizable as a country of its own standing, if not a fully independent one – as demonstrated by Scotland's referendum on independence from the UK in 2014.

Elsewhere, where pairs of would-be nations are present in similar (though admittedly not identical) circumstances, a decision has been made to include one but not the other. So CATALONIA *is in, but Pais Vasco is not. Greenland is in, but Nunavut is not.* RYUKYU *is in, but Ho'aido is not.*

Another would-be country that is not included is the Islamic State of Iraq and Syria, or ISIS. At the time of writing, forces from this proto-state were fighting for control of large swathes of the Middle East. As it happens, this area is roughly coincident with Kurdistan, which declared independence in January 1946 but with little effect. It is impossible to say whether ISIS has a greater chance of enduring as a de facto sovereign entity, or whether it will meet a similar fate to that of AZAWAD, *the Tuareg state in northern Mali that declared independence in 2012 but was dissolved within a year.*

Given this inherent uncertainty of sovereignty in some cases of would-be nation states, a small number of alternative approaches have been included. These cast a critical light on the relationship between territory and sovereignty that holds good for most of the non-countries in these pages. ANTARCTICA *has somehow avoided — or at least put on hold — the process of being carved up into national territories. It signposts one possible version of the future for certain parts of the world. Other, perhaps more fanciful cases, such as* ATLANTIUM *and* ELGALAND-VARGALAND, *indicate unconventional, thought-provoking alternatives. All raise the possibility that countries as we know them are not the only legitimate basis for ordering the planet.*

A FEW WORDS ABOUT BORDERS

James McCarthy removed his sun helmet and scratched his head thoughtfully. His hair was slick with perspiration. The air was sticky and the grove of stunted papaya trees offered little in the way of shade. He was beginning to appreciate the magnitude of his assignment. Appointed by the King of Siam in 1880 to survey his dominion's borders, McCarthy seemed to be the only person interested in the task. Except the king himself, of course. And a few of his courtiers.

Here, local elders told Mr McCarthy, the border with British Burma was defined by two papaya trees, but they couldn't remember which ones. To them it wasn't important. They had left McCarthy in the hot sun to decide for himself.

Nineteenth-century ideas of clearly drawn boundaries were novel in many parts of the world, as they had been once in Europe. At first, the King of Siam found the European obsession with precise borders baffling and rather irritating, but he realized their significance when the British started cutting down swathes of forest and planning railroads. Knowing exactly where his kingdom ended now had a meaning, so he hired Mr McCarthy.

Until McCarthy completed his work, Siamese provinces were not geographically well-described. A province existed in a particular place but the place did not define it. The land itself was almost coincidental. What mattered were the people. And where a boundary did exist, it was seldom a continuous line. It wasn't even a zone. In fact it only occurred where it was needed, such as along a track or pass used by travellers. In other places, where people seldom set foot, there was no point in deciding a boundary. Further, borders between adjacent kingdoms did not necessarily touch, often leaving large unclaimed regions of forest, jungle or mountains. And in practice it was quite possible for towns to have multiple hierarchical relations of authority with more than one ruler and hence — disturbingly for Mr McCarthy — to be part of more than one state.

The conventional political world map we know today has been created thanks to the endeavours of people such as James McCarthy. It is the product of improvements in cartography and surveying and the historical desire of European countries to carve up the world's land surface into colonies. The magnificent cakes they found in Africa and elsewhere.

Borders are important to contemporary countries. They represent the 'skin of the state', the edge of a country's territorial control. In a physical sense they effectively define a modern country, and thus play an important symbolic role in constructing the identity of the state. This is a central reason why many international borders are heavily fortified and closely guarded.

However, depiction of national borders on the world map portrays only a selected version of reality. The confident lines separating fully fledged nation states indicate that there is one universally recognized country that rules over every square centimetre of land. But that isn't quite the case, even without our countries that don't exist.

Many recognized countries have borders that have never been precisely defined and agreed by treaty, a process known as 'delimitation'. And a large number of those that have been delimited have never been actually marked on the ground, or 'demarcated' to use the official term. The lack of delimitation and demarcation frequently spirals into disputes between countries. A classic example occurs in Jammu and Kashmir, where Pakistan, India and China have disagreed for over sixty years on what territory belongs to which country. These and other borders in this Atlas have been drawn with what appears to be a definitive line, but in reality many of these boundaries are far from conclusive.

This is why every United Nations publication containing maps always comes with the somewhat perverse health warning that the maps do not 'imply the expression of an opinion concerning the delimitation of frontiers or boundaries'. The same lack of agreement can also play into the hands of would-be nation states.

Such hazy frontiers also highlight another facet of countries, recognized and otherwise, that remains a given: nothing is set in stone. History is littered with the corpses of would-be states that never made it, empires that dissolved and recognized countries that disappeared into the embraces of more powerful neighbours. Circumstances change. In more recent times there is certainly a tendency for states to endure. In the last fifty years or so, many more new states have been created than have disappeared. Given that knowledge, it seems quite likely that at least some of today's 'countries that don't exist' may one day emerge into the bright light of fully fledged nationhood.

EUR

OPE

ISLE OF MAN

Also known as Ellan Vannin, Mannin

Self-governing dependency of the British Crown, but not part of the United Kingdom or the European Union.

DECLARED:

~

CAPITAL:

Douglas

POPULATION:

85,000

AREA:

572 km²

CONTINENT:

Europe

LANGUAGE:

English, Manx

This is the British Isles, but not as most British people know it. They make their own rules here, and have done for a long time. Their parliament, the Tynwald, is the world's oldest continuous ruling body. It has governed the island since the arrival of the Vikings in the late eighth century. Tynwald means 'assembly field' in Old Norse and for over a thousand years the inhabitants of the Isle of Man have gathered in the same spot, at the summer solstice, to hear the laws of their land proclaimed and to air their grievances. Nowadays, the parliament convenes all year round inside a whitewashed building they call 'the wedding cake', but once a year they like to get outdoors and feel the sun on their faces, to connect with their roots.

They are ancient roots. Before the Norsemen, the island was inhabited by Celts, who left their language, Manx, a sister tongue of Irish and Scots Gaelic. The Vikings ruled the island for nearly five centuries before control passed briefly to the King of Scotland in 1266 and then permanently to the English Crown. The two crowns are now one and the same, and the British monarch is head of state. But being outside the United Kingdom, the island has long set its own taxes. In centuries past, this spawned an interest in smuggling, helped by its geographical position roughly midway between Scotland, England, Wales and Ireland. Today the low taxes support a thriving global financial services industry.

But the island's democratic credentials are second to none. It chalked up a world first in 1881, giving (propertied) women the vote in parliamentary elections. In 2006, it was the first Western European nation to lower its voting age to sixteen. Not bad for a country that doesn't exist.

ISLE OF MAN
Tynwald
Douglas
IRISH SEA
NORWAY
ATLANTIC OCEAN
NORTH SEA
ISLE OF MAN
IRELAND
UNITED KINGDOM
GERMANY
FRANCE

CIRCASSIA

State conquered by Russia in the late nineteenth century.

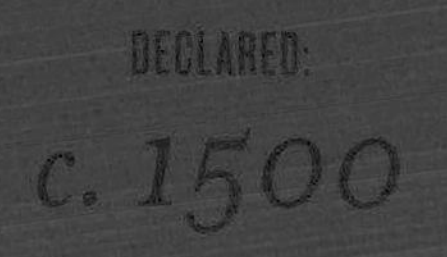

DECLARED:

c. 1500

CAPITAL:

Sochi

POPULATION:

700,000

+ >3 million diaspora

AREA:

33,000 km²

CONTINENT:

Europe

DISSOLVED:

1864

LANGUAGE:

Circassian

Tevik Esenç died a lonely death. He passed away in a village hut with a dirt floor in 1992, the last native speaker of a language called Ubykh. The most striking feature of Ubykh is its large number of consonants: eighty-two in total, with just three vowels – but only recordings remain of this dead tongue, ghostly reminders of its astonishing wealth of sounds.

Tevik Esenç died, as he was born, in exile from his homeland across the Black Sea. The Ubykh lived in Sochi, former Soviet 'worker's resort' and host city of Russia's first winter Olympic Games, but also a place of tragedy and sadness. The Ubykh are one of a patchwork of tribes inhabiting the rugged mountains of the Caucasus, a group of peoples known as the Circassians. Russia conquered Circassia as the Ottoman empire crumbled; vast numbers of Circassians died while many more fled the killing grounds in overcrowded leaking ships, only to end up in camps of misery among the declining Ottomans. The final Circassian warriors met defeat and massacre at Sochi in 1864, a century and a half before those Olympics. What followed is a darker chapter in the region's history: a campaign to empty the territory of Circassians. Some describe it as migration and exile. Others use the word 'genocide'. Perhaps a million Circassians perished.

Forty years after the expulsion of the Ubykh, in 1904, the man destined to be the last speaker of their language was born. Raised by his farmer grandparents, Tevik Esenç served a term as village mayor, rising to a civil service post in Istanbul. With Mr Esenç's death, aged eighty-eight, went the last primary source not only of this Caucasian tongue, but also of Ubykh culture, mythology and customs. The death of an elderly man in Turkey was also the death of a people.

UKRAINE
RUSSIA
CIRCASSIA
Sochi
BLACK SEA
CASPIAN SEA
GEORGIA
Tbilisi
ARMENIA
AZERBAIJAN
Baku
Ankara
TURKEY
SYRIA
CYPRUS
IRAN
LEBANON
Damascus
Baghdad
ISRAEL
WEST BANK
IRAQ
Amman
GAZA
JORDAN
Cairo
EGYPT
SAUDI ARABIA

CHRISTIANIA

Communal self-governing society in Denmark.

DECLARED:

26 September 1971

POPULATION:

850

AREA:

0.34 km²

CONTINENT:

Europe

LANGUAGE:

Danish

It arose as a social experiment in 1971: illegal and anarchic, a paragon of alternative values. A group of hippies, imbued with the 1960s spirit of cultural revolution, began squatting on an abandoned government site. The operation started with perfect symbolism, in a former military barracks right in the heart of Denmark's capital city, Copenhagen.

On 26 September, the squatters founded what they called the Freetown of Christiania. This was a community set up in opposition to Danish society, a self-governing collective where everyone took responsibility for the well-being of the entire community. Decisions were taken by 'direct democracy', a consensus in which everybody had to agree. Within a year, the defence ministry had granted Christiania's citizens the collective rights to use the site, so long as they paid for their water and electricity.

Although inherently rebellious, Christiania subsequently went through what in other circumstances might be called a rebellious phase. Among the economic activities that flourished in the new society was a trade in hard drugs. Decked in psychedelic murals, Pusher Street became synonymous with heroin, cocaine and amphetamines. Eventually, a drugs-related murder and a series of overdoses led to Christiania banning the hard stuff, leaving only cannabis openly available for residents to enjoy. This was a carefree community, but with certain limits.

The government of Denmark has never been quite sure what to do about the hippy haven on its doorstep. Christiania openly defies laws that apply to the rest of Danish society and has been allowed to get away with it. Until 2012, when the government tested the community's values, offering the land for sale to them at a big discount. Although a victory in one sense, this was also capitalism's revenge. A self-owned Christiania would mean violating their basic principles around property ownership. The collective has until 2018 to buy its own future.

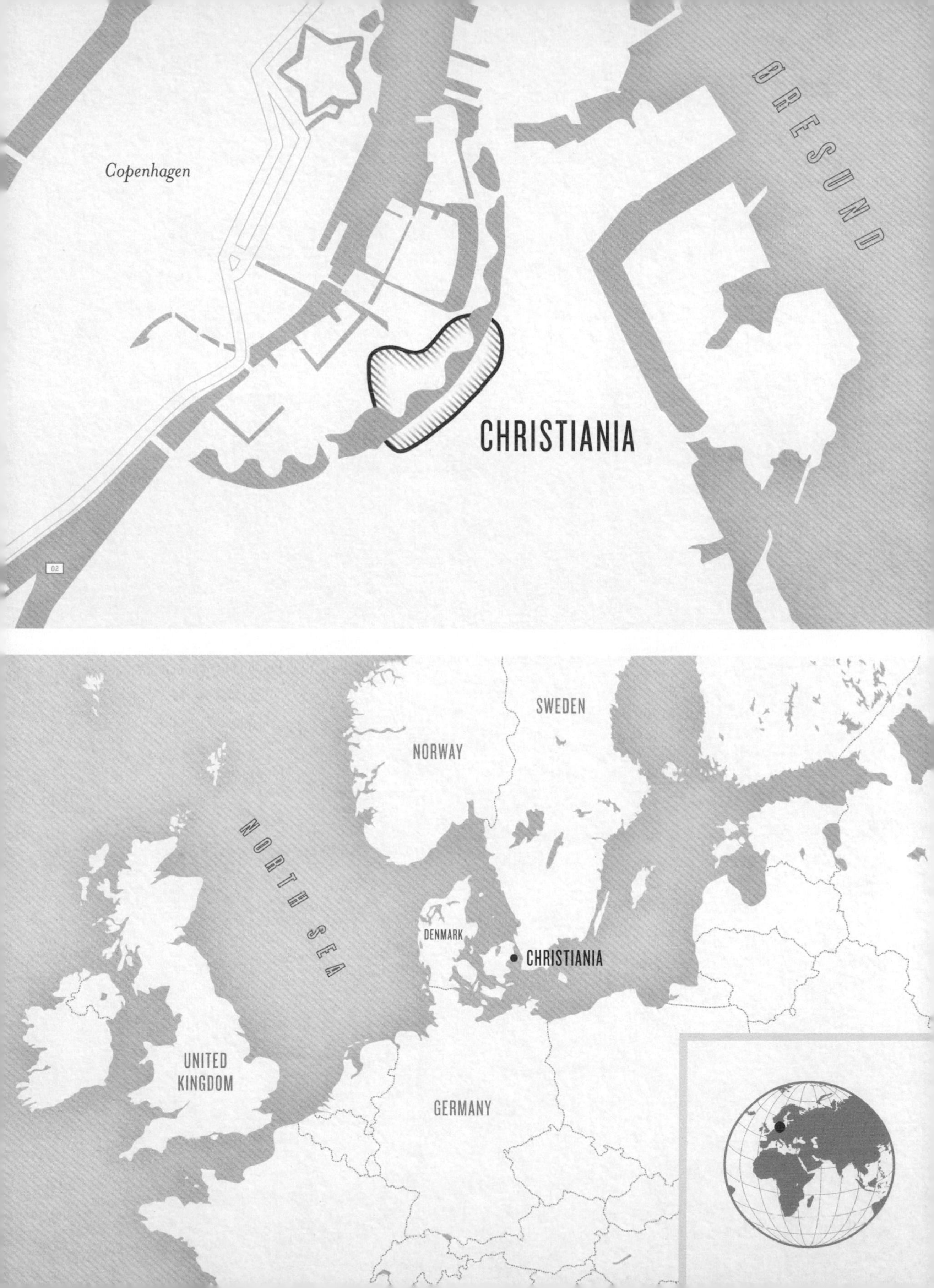
Copenhagen
ØRESUND
CHRISTIANIA
SWEDEN
NORWAY
NORTH SEA
DENMARK
CHRISTIANIA
UNITED KINGDOM
GERMANY

NORTHERN CYPRUS

North-eastern third of the island declared independent after invasion by Turkey.

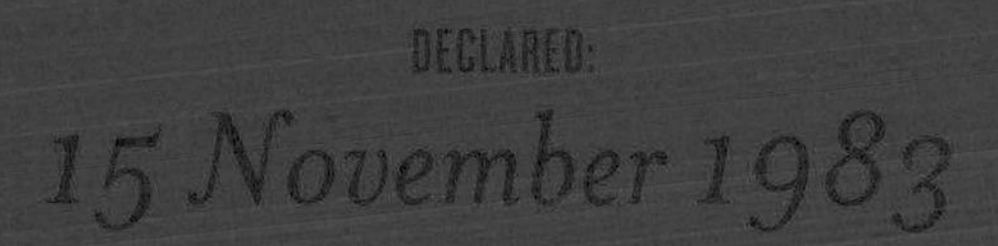

DECLARED:

15 November 1983

CAPITAL:

North Nicosia

POPULATION:

294,906

AREA:

3.355 km²

CONTINENT:

Europe

LANGUAGE:

Turkish

Varosha is a forbidden zone. The signs by the barbed-wire fence say so in five different languages. This has been a no-go area since the first paratroopers drifted down out of the early morning sunshine in the summer of 1974.

As the Turkish soldiers approached, residents of this Greek-Cypriot seaside town fled in their bell-bottom trousers, hastily packed bags under their arms. A mother held a child still blissfully asleep in a blanket. An elderly lady, dressed entirely in black, hobbled slowly with her walking stick as sporadic gunshots punctuated the crackling of automatic fire.

They left behind a thriving Mediterranean resort, the best beach on Cyprus, a glitzy playground for the rich and famous. After years of feuding between the island's two communities led to a coup d'état inspired by Greece's ruling military junta, Turkish troops occupied the north-eastern part of the island. One side talked of the liberation by Turkey, the other called it an invasion.

The inhabitants of Varosha still have hopes of one day returning home. Over the passing decades, both sides of the Cypriot ethnic divide have haggled over many issues, not least the ownership of land. Title deeds say the land was originally owned by Ottoman Turkish charitable foundations, but when Cyprus became a British crown colony in 1925, the Brits began selling parcels of Varosha real estate. Most were bought by Greek entrepreneurs.

Meanwhile, Varosha has become a ghost town, a time warp encased in dust. Through the fences, beyond the thicket of signs, abandoned hotels sit ragged and skeletal. Prickly pear bushes thrust through apartment walls and weeds rampage where the jet set once strolled. A car dealership still displays its grimy stock of 1974 models, while pockmarked mannequins dressed in outmoded fashions stare blankly from wrecked shop windows. But no one is sunbathing on Varosha's golden sands.

TURKEY
NORTHERN CYPRUS
North Nicosia
Varosha
CYPRUS
BUFFER ZONE
SYRIA
LEBANON
Beirut
Damascus
MEDITERRANEAN SEA
WEST BANK
Amman
GAZA
ISRAEL
SUEZ CANAL
JORDAN
EGYPT
SAUDI ARABIA

MORESNET

Created in 1816 around a strategic zinc mine when Prussia, the Netherlands and later Belgium could not agree on sovereignty. Renamed Amikejo, it became a self-determining global haven for Esperantists before its disappearance after World War I with the Treaty of Versailles.

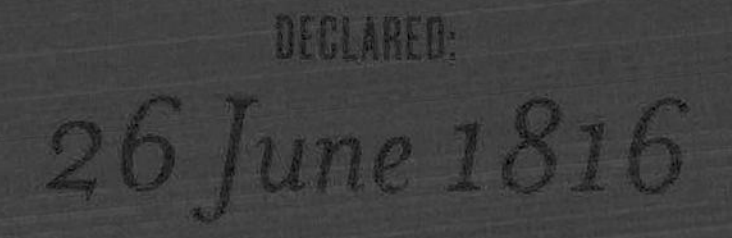

DECLARED:

26 June 1816

CAPITAL:	POPULATION:
Kelmis	*3,000*
AREA:	CONTINENT:
3.4 km^2	*Europe*
DISSOLVED:	LANGUAGE:
27 June 1915	*German, French, Esperanto*

0 0.5 1 2 3 4 5 KILOMETRES

The settlers entered Moresnet from the north, at a spot not far from the railway track running between Aachen and Welkenraedt. Past the three granite columns marking the highest point in the Netherlands, they strode down through the thick forest into what was once one of Europe's richest mining valleys.

In its nineteenth-century heyday, this was a thriving settlement: its schoolhouse resplendent, its workers well housed. Free of outside restrictions, the mining company organized health insurance and interest-bearing savings accounts. But by the dawn of the twentieth century the zinc was all but gone. Gone too was an attempt to transform Moresnet into a giant casino. In 1903, when roulette tables were outlawed in neighbouring Belgium, a businessman made an offer to the town's burgomaster and councillors: a million francs a year to open their doors to gamblers. These golden, spinning-table dreams did not come to pass. Disaffected youths turned to crime and smuggling to make ends meet.

In August 1908, people gathered in the pavilion of the shooting association, beneath the grimy smokestacks, to join in a new beginning. Their spiritual leader was Dr Wilhelm Molly, a Moresnet general practitioner and founder of the local postal service. Spirits were high, stirring speeches were delivered and the miners' brass band struck up during the intervals. Dr Molly's dream of an independent Esperanto city-state was proclaimed. They renamed Moresnet in the vernacular, henceforth to be known as Amikejo, the 'place of friendship'. This territory would be a global haven, a sanctuary from 'all that is absurd and unworthy in convention, all that the ignorant centuries have imposed upon us'.

A coat of arms was solemnly unveiled. The band played the Amikejo March, composed for the ceremony and sung to the tune of 'O Tannenbaum', the Christmas tree song. It was to be their national anthem.

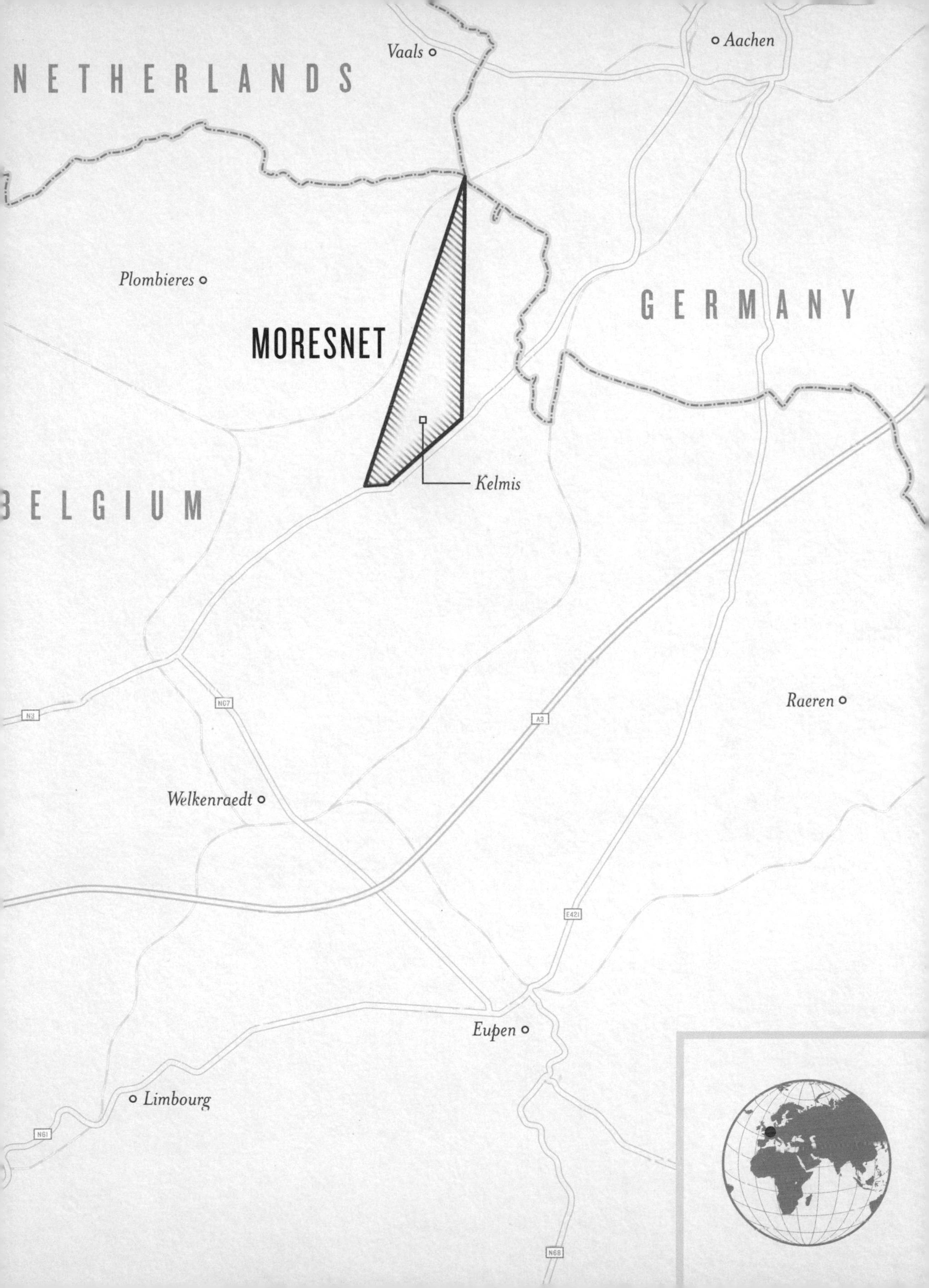

Vaals
Aachen
NETHERLANDS
Plombieres
GERMANY
MORESNET
Kelmis
BELGIUM
N3
N67
A3
Raeren
Welkenraedt
E421
Eupen
Limbourg
N61
N68

FORVIK

Also known as Forewick Holm

Shetland island state created by an English yachtsman.

DECLARED:

23 February 2011

POPULATION:

1

AREA:

0.01 km²

CONTINENT:

Europe

FOUNDER:

Stuart Hill

LANGUAGE:

English

Stuart Hill spent his first night in the North Sea archipelago of Shetland in hospital suffering from hypothermia. He had been rescued by coastguards after capsizing in his converted rowing boat, MAXIMUM EXPOSURE, *during an ill-fated attempt to circumnavigate the British Isles. It was the eighth and final time he was rescued in the four months since leaving Kent in England, and his maritime mishaps had earned him a nickname in the British media. They called him Captain Calamity.*

But he liked Shetland, so Hill decided to stay. By 2008 he had taken up residence on the windswept islet of Forvik, and began a campaign for self-determination based on Shetland's cultural affinity with Scandinavia, a facet of its Viking heritage. The islands, he argued, technically remained part of the Norse empire that had passed into history. Hoping to be the vanguard of a wider Shetland autonomy, Mr Hill posed a query: could the UK government explain the basis for their perceived authority in Shetland? The answer was a deafening silence.

Nobody took much notice until Hill was sued by a debt agency. At last he saw an opportunity to test his legal argument that Scotland had no jurisdiction over Shetland. Standing in the dock, his exchange with Sheriff Philip Mann was genial but ultimately fruitless. The sheriff said he could not agree with Mr Hill's position. 'If you're correct I might as well just fold up my papers and walk out now,' he told him. 'I don't see how I can competently make a ruling that I'm sitting here incompetently.'

When asked for his position on the case itself, Mr Hill offered no defence and said he would appeal to a higher court. Sheriff Mann made him bankrupt but wished Captain Calamity the best of luck.

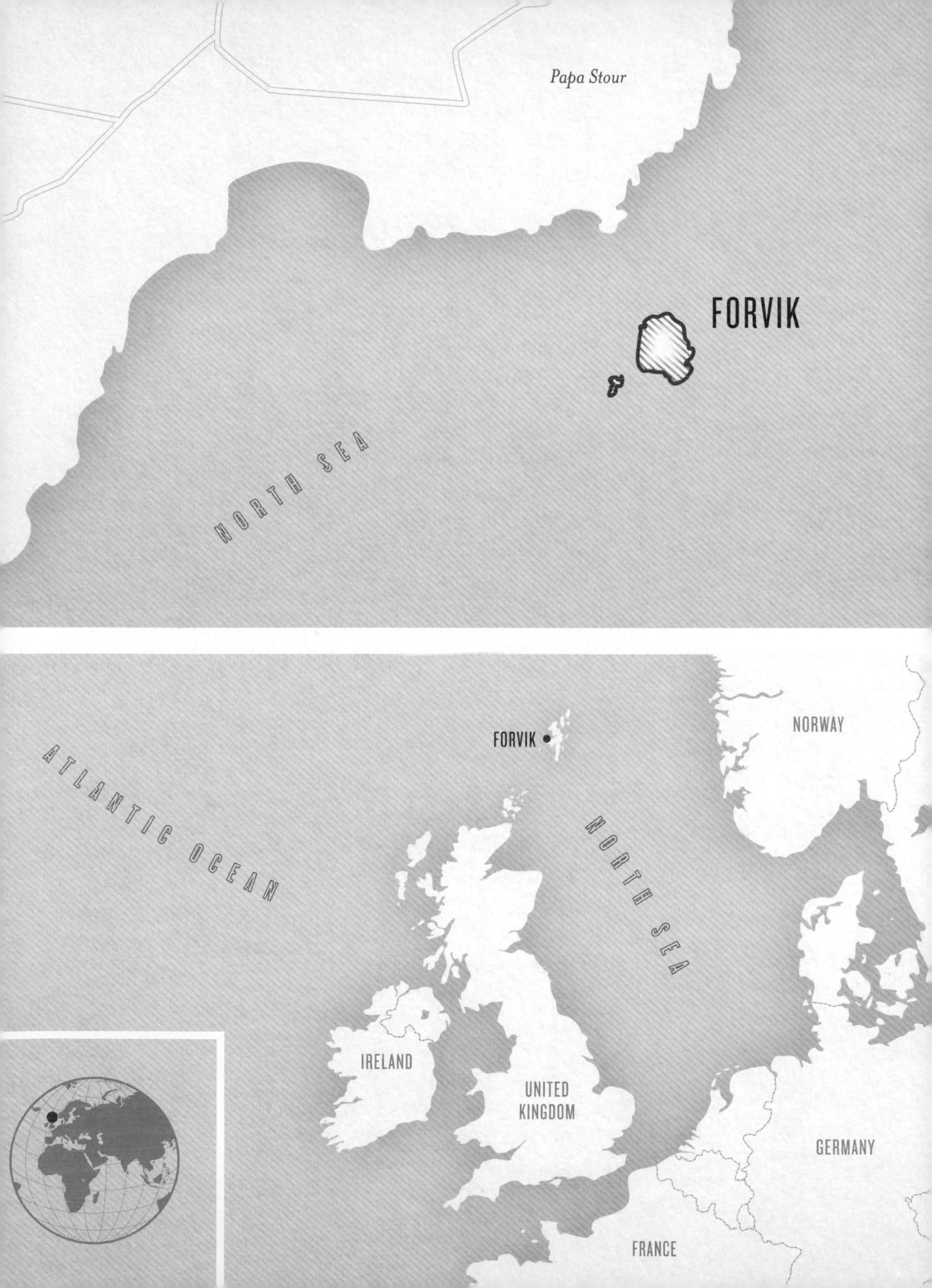
Papa Stour
FORVIK
NORTH SEA
FORVIK
NORWAY
ATLANTIC OCEAN
NORTH SEA
IRELAND
UNITED KINGDOM
GERMANY
FRANCE

ABKHAZIA

Breakaway enclave of Georgia supported by Russia.

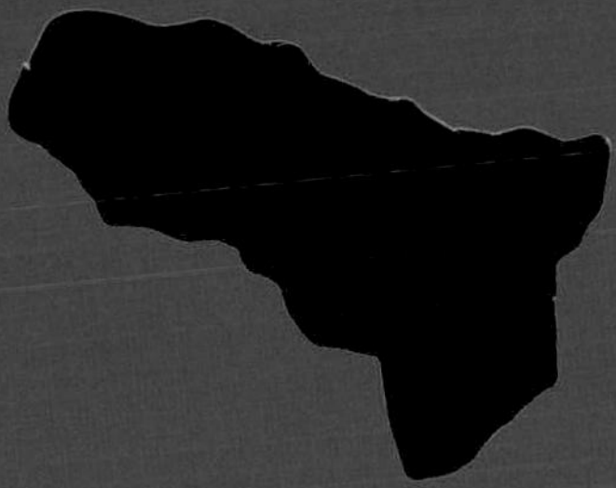

DECLARED:

8 March 1918; 31 March 1921; 23 July 1992

CAPITAL:

Sukhumi

POPULATION:

240,000

AREA:

8,660 km²

CONTINENT:

Europe

LANGUAGE:

Abkhaz, Russian

They came to burn down the National Library one balmy afternoon in August 1992. Local residents helped the librarians douse the flames, carrying buckets of seawater from the beach. Two months later, when skirmishes were routine on the streets of Sukhumi, they tried again, at the Abkhaz Research Institute of History, Language and Literature. This time the Georgian paramilitaries were more thorough. They used kerosene and kept civilians back at gunpoint. A fire engine arrived and was turned away. In the morning, when the academicians arrived for work, most of what they found was ashes.

The Abkhaz and the Georgians have different ethnic roots, although they shared this stretch of coast for centuries. After two brief attempts at independence, Abkhazia was sucked into Georgia and became part of the Soviet Union. The situation deteriorated under Soviet tyrant Joseph Stalin, an ethnic Georgian himself. Abkhazia's status was downgraded, Abkhaz schools were closed and their language banned. Georgian peasants were resettled in Abkhazia and the Abkhaz became a minority in their own homeland.

The city of Sukhumi was transformed into a favourite holiday spot for Stalin's elite. Government sanatoria and guesthouses for the KGB – the Soviet security service – sprang up along the subtropical coastline. Stalin kept his own personal dacha, perched on a hilltop with spectacular views across tangerine groves to the sparkling Black Sea. Peacocks strutted the grounds of the municipal spa. Communist Party bigwigs and senior KGB officers strolled the palm-lined boulevards and supped the finest Georgian brandy.

When the Soviet Union disintegrated, an ethnic time-bomb was detonated. Afraid of being completely swamped, the Abkhaz declared their own republic. Georgian troops occupied Sukhumi and books were burned. So began a bloody, year-long war of secession that ended with Russian troops keeping the peace and propping up a barely autonomous Abkhazia.

RUSSIA
ABKHAZIA
Sukhumi
BLACK SEA
GEORGIA
Tbilisi
ARMENIA
AZERBAIJAN
TURKEY
IRAN
IRAQ
SYRIA

CATALONIA

A corner of Spain that seeks full statehood.

DECLARED:

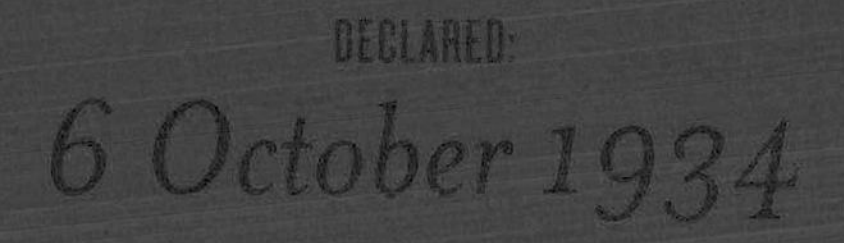

6 October 1934

CAPITAL:

Barcelona

POPULATION:

7,600,000

AREA:

32,114 km²

CONTINENT:

Europe

LANGUAGE:

Catalan, Spanish

0 50 100 200 300 KILOMETRES

More than two million people turned out to vote in Catalonia in what was billed as a 'consultation of citizens'. In the Spanish capital, Madrid, they saw it differently. Spain's highest court, the Tribunal Constitucional, ruled the whole thing unlawful. Not only was the non-binding referendum on independence illegal but also the law passed by the Catalan regional government allowing it to arrange such votes.

Even within Catalonia the situation was not clear-cut. Certainly over two million people had voted in 2014, and a majority of them had said yes to independence. But fewer than half of those eligible to cast a ballot had done so. The head of the Catalan regional government called the referendum a 'total success'. In Madrid, the Spanish justice minister dismissed the exercise as 'sterile and useless'.

None of which will make the Catalan question go away. With their own ancient language and a distinct history stretching back into the Middle Ages, a sizeable number of Catalans consider themselves a nation that is separate from the rest of Spain. That detachment is still fuelled by bitter memories of General Franco's thirty-six-year dictatorship, in which Catalonia endured a cruel and systematic attempt to obliterate her culture.

This north-eastern region of Spain has enjoyed the hallmarks of independence before. They had a constitution until the eighteenth century, and their own parliament, twice, in the twentieth. They even managed to proclaim Catalan independence in October 1934, but it lasted just ten hours before the Spanish army restored order. Within a few years, General Franco had taken over.

Currently, with the return of democracy confirmed under the Spanish Constitution of 1978, Catalonia enjoys a high level of self-government. It runs its own schools, police force, health care system and cultural institutions. But for some Catalans at least, this is not enough.

FRANCE
Marseille
PYRENEES
ANDORRA
CATALONIA
Barcelona
SPAIN
Menorca
Valencia
Majorca
Ibiza
MEDITERRANEAN SEA
Algiers
ALGERIA
MOROCCO

SEBORGA

Principality declared independent from Italy after a referendum in 1995.

DECLARED:

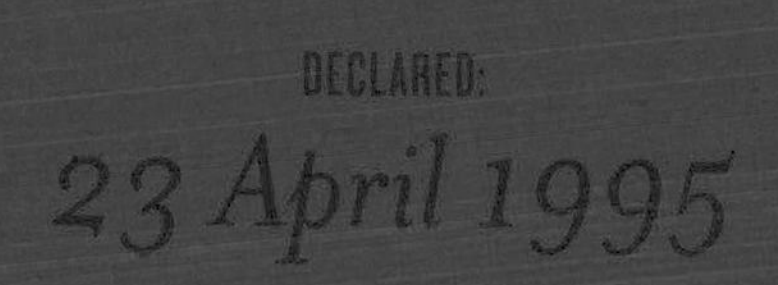

23 April 1995

CAPITAL:

Seborga

POPULATION:

312

AREA:

0.5 km^2

CONTINENT:

Europe

LANGUAGE:

Italian

The head of the local flower-growers cooperative is a placid figure. Giorgio Carbone is an avid smoker with a rugged face and a big black beard. He grows mimosa flowers and lives in a small hilltop town in north-western Italy, an ancient jumble of narrow cobbled streets, wooden shutters and wrought-iron balconies.

Mr Carbone spends many hours in state and church archives painstakingly reconstructing the town's thousand years of history. In 1079, Seborga is designated a principality of the Holy Roman Empire. It remains independent for more than six centuries, until it is sold to the House of Savoy, a transaction that is not registered. This error is subsequently compounded. When the great powers of Europe meet to settle the boundaries of the continent at the Congress of Vienna in 1815, Seborga is not mentioned. When the many small states on the Italian peninsula are unified to form Italy in 1861, Seborga is not mentioned. In 1946, after the abdication of the last Savoy king, Victor Emmanuel II, Italy becomes a republic, but Seborga is not mentioned.

In 1995, Mr Carbone puts it to the good people of Seborga that their town is not, after all, a part of Italy. A local referendum confirms this stance and ratifies Seborga's independence. The former head of the flower-growers cooperative accepts the honorific title of His Tremendousness, elected prince for life. Seborga's sovereign has been elected since the Middle Ages, so this is a return to tradition. Their kindly ruler holds court at the Bianca Azzura bar, often wearing light blue sash, sword and rosette medallions. He travels in a black Mercedes with a Seborgan licence plate: number 0001. His Tremendousness passed away in 2009, but his loyal subjects carry on. They still pay taxes to Italy, but that state refuses to recognize the independent Principality of Seborga.

San Martino
SEBORGA
ITALY
< MONACO
Madonna della Neve
FRANCE
MONACO
SEBORGA
ITALY
Corsica
Sardinia

TRANSNISTRIA

Separatist region of Moldova, often portrayed in the West as a hotbed of crime and Stalinism.

DECLARED:

2 September 1990

CAPITAL:

Tiraspol

POPULATION:

518,700

AREA:

4,163 km²

CONTINENT:

Europe

LANGUAGE:

Russian, Romanian-Moldavan

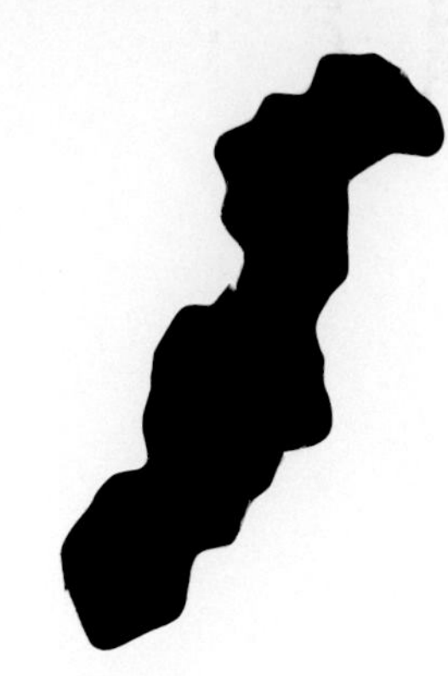

0 50 100 200 KILOMETRES

From Brussels, the EU bureaucrats see it as a black hole in Europe, a hub for money-laundering, people-trafficking and the illegal arms trade that seethes with epic tales of the criminal underworld, all beyond the reach of international law. Politically, they spy a time warp back to the days of the old Soviet Union, a place where the hammer and sickle continues to flutter in the corner of the national flag. A hawk-eyed statue of Lenin still graces the front of the parliament building in Tiraspol.

Since separating from its neighbour and foe across the river – Moldova – this sliver of land on the left bank has been bolstered by Russia and a sense of collective victimization among its residents. More than 300 miles from the nearest Russian border, they see themselves as Russian people marooned by the collapse of the USSR. Back in the old days, fresh produce from 'the Kremlin Garden' north of Tiraspol was flown direct to Moscow every day. No surprise that outsiders simultaneously observe a Stalinist backwater and a criminal Ruritania.

Yet order is maintained, thanks in large part to a shadowy public body known only as 'the Sheriff'. Blurring the line between business and politics, the Sheriff steers a course between chaos and prosperity, navigating the passage from state socialism to state capitalism. It is a private enterprise with political clout. Some call it the economic arm of the state, owning petrol stations and supermarket chains, a mobile phone network and the country's leading football club, FC Sheriff Tiraspol. To the uninitiated, Sheriff and Transnistria appear as one. It is the largest employer and the Sheriff's badges appear on almost every building. The name reflects the previous occupation of its two founders, former agents of state security. As its website proudly declares, the Sheriff is 'Always with you!'

CARPATHIAN MOUNTAINS
UKRAINE
TRANSNISTRIA
MOLDOVA
Chisinau
Tiraspol
Odessa
ROMANIA
DANUBE
Sevastopol
Bucharest
BLACK SEA
Varna
BULGARIA
Istanbul
TURKEY

RUTHENIA

Republic for a day in March 1939.

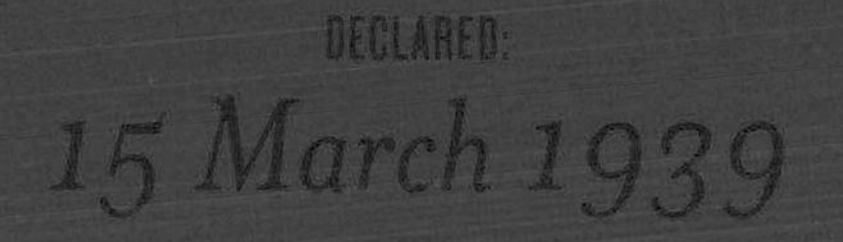

DECLARED:

15 March 1939

CAPITAL:	POPULATION:
Khust	*c. 1,000,000*
AREA:	**CONTINENT:**
12,777 km²	*Europe*
DISSOLVED:	**LANGUAGE:**
16 March 1939	*Ruthenian, Ukrainian*

The Battle of Krasne Pole was a short-lived encounter, a David and Goliath affair, only without the counter-intuitive ending. Ruthenian forces took up their position on a plain on the north side of the Tysa River, a tributary of the mighty Danube. They would defend their fledgling republic, against the might of the Hungarian army, at the strategic Veriatsk Bridge.

The defenders were a hotch-potch of poorly equipped patriots, joined by school students and their teacher. A few sat expectantly in the snow behind Maxim machine guns; most were armed with hunting rifles. Across the river lay a vastly superior Hungarian force, supported by tanks, armoured vehicles and an air force. This was the first resistance they had met since crossing the unguarded border, under authorization from Germany, invading Ruthenia.

It was 15 March 1939. Beneath the red volcanic hills of the Carpathians, the shooting began. Some claim these as the first shots of World War II.

Only that morning, a dozen kilometres upstream in the town of Khust, a new republic had been created. Its president was a Greek Catholic bishop and man of letters. Among his works, a practical grammar of the Ruthenian language. He was an apt guardian of Ruthenian culture and identity. Independence was declared. Men in dark suits and stiff white collars applauded the announcement from wooden benches. At sunrise, they had been part of Czechoslovakia. By lunchtime they had their own nation-state. By evening, they had disappeared into the clutches of a budding Nazi empire.

In 1945, Ruthenia's president for a day was arrested by the Soviet secret police, and died soon after in a Moscow jail. His short-lived republic vanished into westernmost Ukraine, but the ghost of independence lives on. More than half a century later, revivalists once more demand Ruthenian self-determination.

POLAND
UKRAINE
SLOVAKIA
Kosice
CARPATHIAN MOUNTAINS
Khust
RUTHENIA
HUNGARY
Cluj-Napoca
ROMANIA
SERBIA
DANUBE
Belgrade
Bucharest
BULGARIA

SEALAND

Founded in 1967 on the high seas about 7 nautical miles east of the UK.

DECLARED:

2 September 1967

CAPITAL:	POPULATION:
Sealand	*27*
AREA:	CONTINENT:
4,000 m²	*Europe*
FOUNDER:	LANGUAGE:
Roy Bates	*English*

Joan Bates and her son Michael are afraid, but they know that if they keep their nerve the men in the ship below, bobbing rhythmically on the North Sea swell, cannot board their island fortress. It is night-time and it is cold. It is always cold on the bleak, 20-metre-high platform. Sometimes Joan is so cold she feels faint while hanging out the washing.

Built in 1943 by the Royal Navy, the concrete-and-steel platform was equipped with anti-aircraft guns to shoot down Luftwaffe planes. Abandoned after the war, it eventually came to the attention of Joan's husband Roy, a former infantry major and owner of a chain of butchers' shops, who was looking for a new challenge. Roy quickly took up residence to broadcast pop music, a classic Sixties 'pirate radio station' lying just outside British territorial waters, safe from prosecution.

In time a court case arose nonetheless, with Joan's son, Michael, the accused. When an auxiliary vessel working on a nearby buoy sailed too close, foul language bloomed in the cold air. The situation escalated, and Michael fired warning gunshots across their bow. The vessel raced away, back to the Thames estuary, but the next time Michael set foot on the mainland he was arrested and put on trial for firearms offences. After much deliberation, the case was dismissed. In summing up, the judge described it as 'a swashbuckling incident perhaps more akin to the time of Sir Francis Drake', but noted that his court had no jurisdiction because Sealand lies outside British territorial waters.

The Bates family returned to their island stronghold, content with what they saw as Sealand's first de facto recognition. Roy, Prince of Sealand, passed away in 2012, but his legacy reached a fourth generation of Sealand royalty two years later with the birth of Prince Freddy, a grandson for Michael.

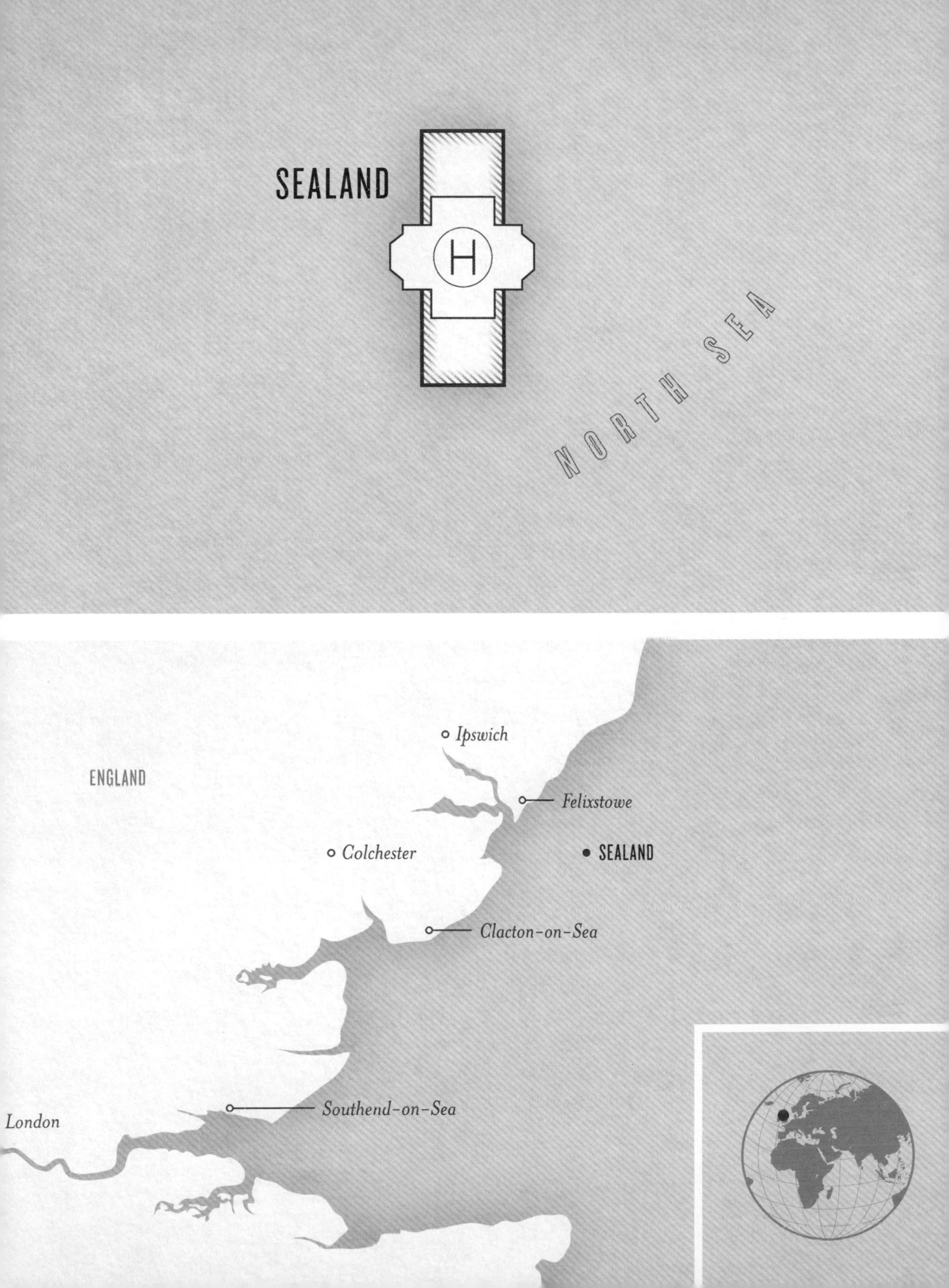

SEALAND
H
NORTH SEA
Ipswich
ENGLAND
Felixstowe
Colchester
SEALAND
Clacton-on-Sea
Southend-on-Sea
London

CRIMEA

Peninsula where independence has been declared four times in a century.

DECLARED:

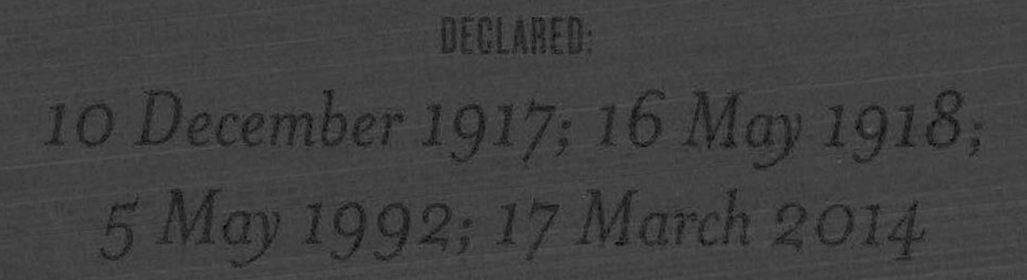

10 December 1917; 16 May 1918; 5 May 1992; 17 March 2014

CAPITAL:
Sevastopol

POPULATION:
2,000,000

AREA:
26,200 km²

CONTINENT:
Europe

DISSOLVED:
10 May 1992; 18 March 2014

LANGUAGE:
Russian, Ukrainian, Crimean Tatar

Margarita Pobudilova emerges from the high-rise office block and shuffles past a long queue of fellow citizens, anxiously awaiting news about their savings. For Mrs Pobudilova, a retired factory worker, the wait continues. Her money will not be returned today.

She invested her life savings in a one-month bond in February but by the time her investment matured, the Crimean peninsula had declared independence from Ukraine and been annexed by the Russian Federation. Ukraine's currency was banned and the Ukrainian bank holding Mrs Pobudilova's deposit was forced to close. Mrs Pobudilova's money has not been accessible since.

History repeats itself. In the late eighteenth century, after the Russo-Turkish War, Crimeans enjoyed nominal independence until their peninsula was annexed by Catherine the Great. In acquiring a warm water port for her navy, she rode roughshod over the people who had controlled Crimea for 500 years: the Crimean Tatars.

Originally from Central Asia, the Crimean Tatars have suffered greatly at the hands of Russia. Crushed by the Red Army in 1917 and 1918, in 1944 Stalin accused the entire Tatar race of treason and deported them back to Central Asia in cattle cars. He settled Russians on their land instead.

By the time those Crimean Tatars still surviving were allowed home after forty-five years in exile, Crimea was predominantly Russian and had been handed to Ukraine as a gift. After the Soviet Union disintegrated, in 1992 Crimea's Russian leaders announced their separation from Ukraine, but changed their minds five days later. In 2014, the same decision was met with the same Ukrainian disapproval. This time, their secession lasted just a day, but with a different ending. Crimea was absorbed back into Mother Russia.

All of which cuts little ice with Mrs Pobudilova and others. They just want access to their bank accounts.

UKRAINE
MOLDOVA
DNIEPER
RUSSIA
ROMANIA
Sevastopol
CRIMEA
BLACK SEA
GARIA
GEORGIA
Ankara
TURKEY
CYPRUS
SYRIA
LEBANON
Damascus
IRAQ
ISRAEL

AFR

ICA

SOMALILAND

Independence declared in 1991 with the boundaries of the former British Somaliland Protectorate.

DECLARED:

26 June 1960; 18 May 1991

CAPITAL:
Hargeisa

POPULATION:
3,500,000

AREA:
137,600 km²

CONTINENT:
Africa

DISSOLVED:
1 July 1960

LANGUAGE:
Somali, Arabic

Sitting cross-legged on a coarse camel-hair carpet, Sayyid Mohamed Abdille Hassan was writing a letter to the grand foreign power attempting to colonize his homeland. It was 1904 and his Dervish army had been waging a holy war against the British for three years. A master of guerrilla tactics, from the saddle of his favourite horse — SOUND OF FLYING GRAVEL — *he tormented a force that was larger and better equipped. He drew the seasoned British troops into the arid furnace of Somaliland, terrain that was familiar to the Dervishes.*

The struggle had been bloody but inconclusive, so Sayyid Mohamed changed tactics. An eloquent poet as well as a visionary leader, he poured his soul into lyrical prose.

I HAVE NO CULTIVATED FIELDS, NO SILVER OR GOLD … IF YOU WANT WOOD AND STONE, YOU CAN GET THEM IN PLENTY. THERE ARE ALSO MANY ANT-HEAPS. ALL YOU CAN GET FROM ME IS WAR — NOTHING ELSE. IF YOU WISH FOR WAR, I AM HAPPY. BUT IF YOU WISH FOR PEACE, GO AWAY FROM MY COUNTRY BACK TO YOUR OWN.

The invaders did not leave what they called the British Somaliland Protectorate. The man they dubbed the 'Mad Mullah of Somaliland' continued his stubborn defiance for another decade and more before succumbing to Unit Z of the newly formed RAF — one of the first uses of air power to crush an insurgency.

It was not until 1960 that the infidel usurpers finally left Somali territory. British Somaliland became independent for five days before joining the former Italian Somaliland to create the Somali Republic. Sayyid Mohamed's poetry is still taught to Somali children, but his dream of national unity was not to be. Persecuted by southerners, Somaliland seceded after a civil war, reverting in 1991 to the boundaries drawn by the infidels.

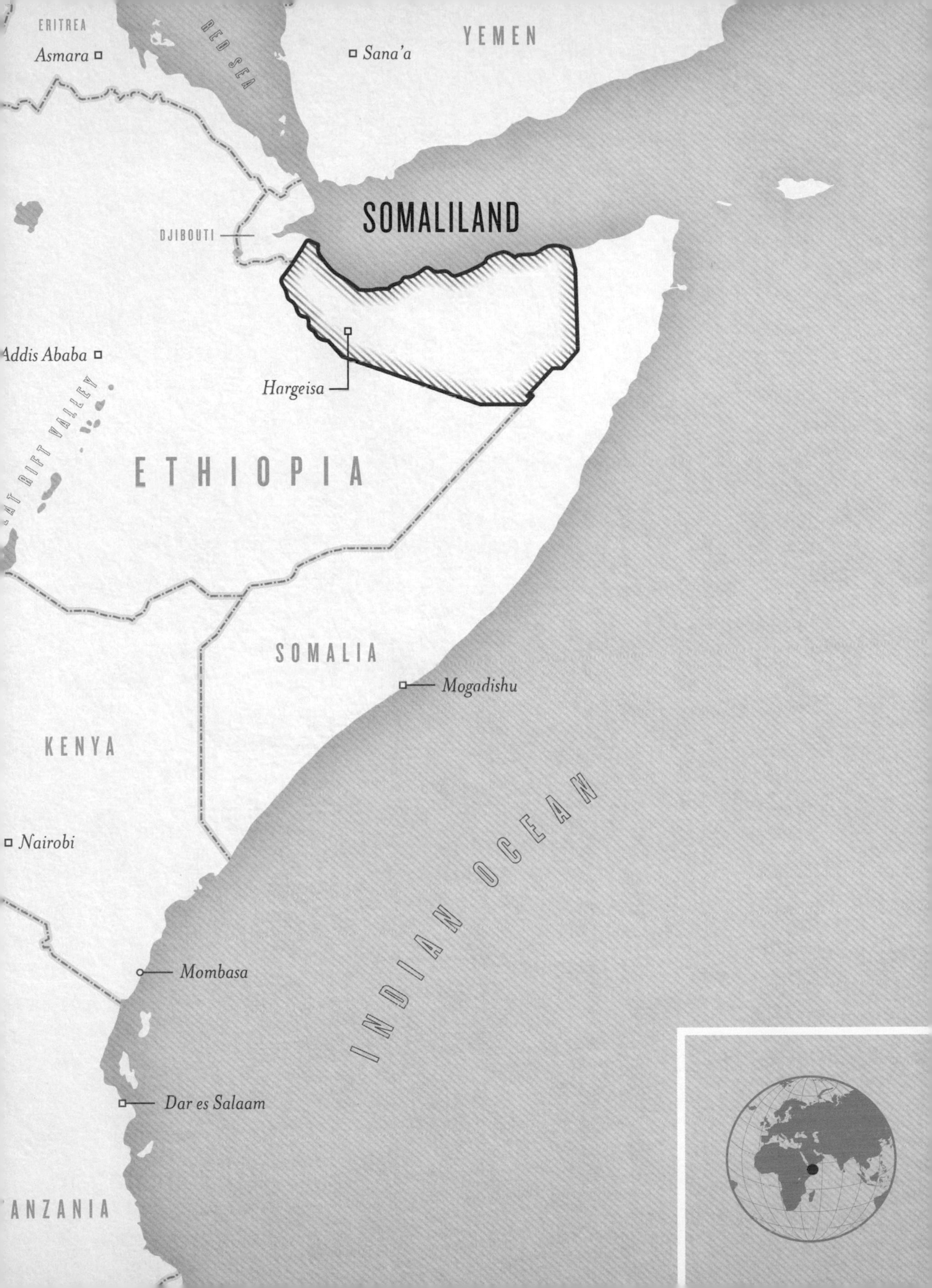

ERITREA
Asmara
RED SEA
Sana'a
YEMEN
SOMALILAND
DJIBOUTI
Addis Ababa
Hargeisa
AT RIFT VALLEY
ETHIOPIA
SOMALIA
Mogadishu
KENYA
Nairobi
INDIAN OCEAN
Mombasa
Dar es Salaam
ANZANIA

MAYOTTE

Indian Ocean island, claimed by Comoros, that chose to forgo independence from France in 1975.

DECLARED:

6 July 1975

CAPITAL:

Mamoudzou

POPULATION:

212,645

AREA:

374 km²

CONTINENT:

Africa

LANGUAGE:

French, Shimaore, Kibushi

0 5 10 20 30 40 KILOMETRES

Self-determination was not supposed to work like this. In its declaration on decolonization, the United Nations was clear on the matter. It solemnly proclaimed the need for an unconditional end to colonialism in all its forms. In the latter half of the twentieth century, the four islands in the Comoros archipelago joined the march towards African independence, though not all marched at the same speed. In a referendum in 1974, an overwhelming majority of votes were in favour of independence from French rule. But nearly all the 'no' votes were cast on the island of Mayotte, where a majority wanted to remain part of France.

After more than a century of European colonial rule, the Comoros declared independence in 1975 for the whole archipelago, including Mayotte. The UN granted the new country membership; all four islands that is. To do otherwise would have threatened the new state's territorial integrity. This well-established principle was seen as a stabilizing influence during periods of decolonization: each new state would inherit the boundaries of colonies on independence. To allow otherwise would be an invitation to fragmentation and potential chaos.

France duly granted independence to three of the four islands – but retained Mayotte. Faced with international disapproval, another referendum was organized and again Mayotte voted to stay French. There was no doubt, it seemed. The people of Mayotte preferred to be marginalized by a European capital 8,000 kilometres away rather than by their own kinsmen on the next island. The UN dismissed the referenda as contrary to international law. France stood fast.

The stalemate continued, annually in the UN chamber, as Mayotte consolidated her position. She worked her way up through the administrative categories to become a fully fledged département of France. In 2014, the island became an official part of the European Union. Self-determination in this case has produced a very postmodern colony.

Mamoudzou
MAYOTTE
MOZAMBIQUE CHANNEL

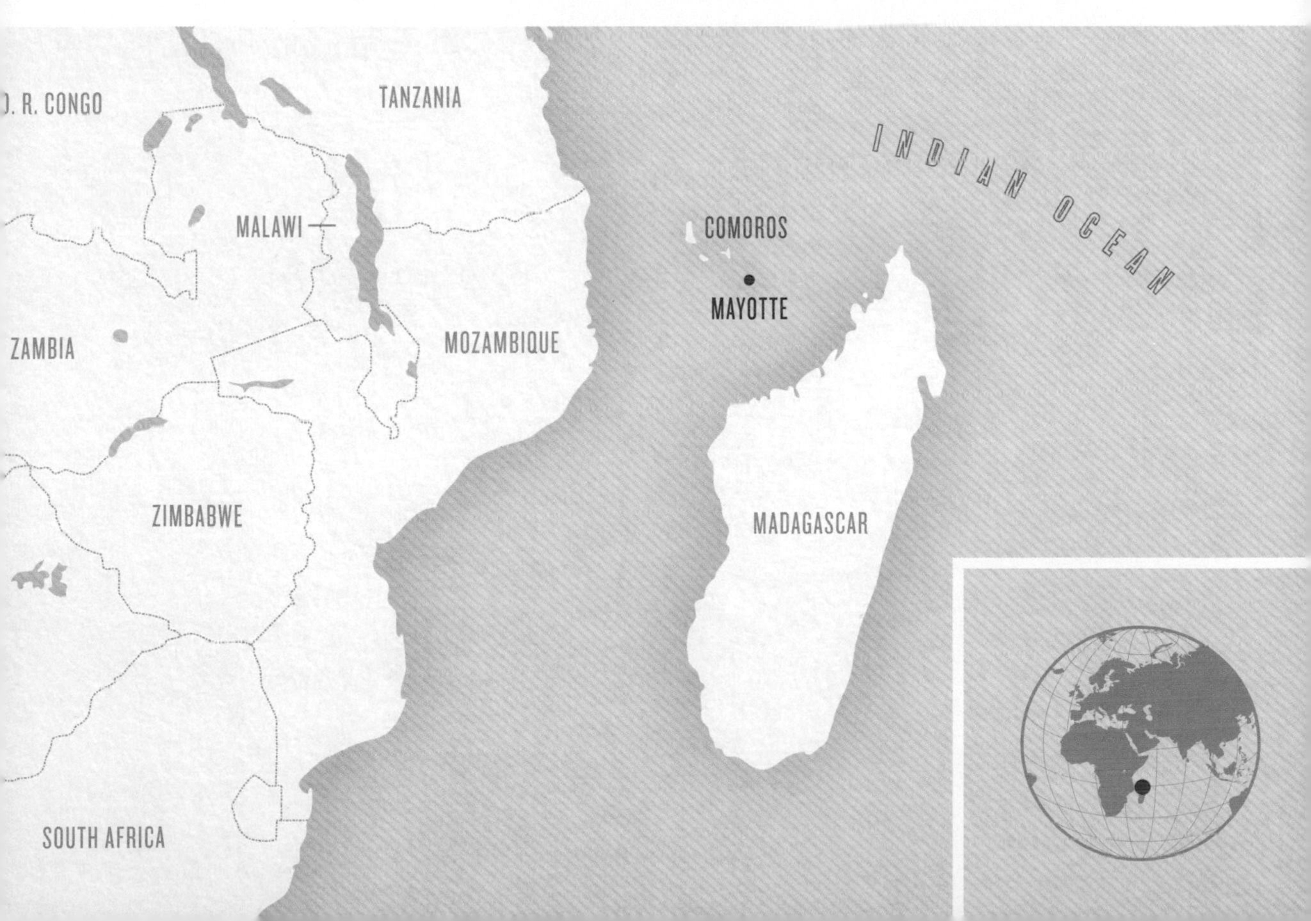
TANZANIA
INDIAN OCEAN
COMOROS
MALAWI
MAYOTTE
ZAMBIA
MOZAMBIQUE
ZIMBABWE
MADAGASCAR
SOUTH AFRICA

AZAWAD

Saharan republic for less than a year.

DECLARED:

6 April 2012

CAPITAL:	POPULATION:
Gao	*1,300,000*
AREA:	CONTINENT:
820,000 km²	*Africa*
DISSOLVED:	LANGUAGE:
14 February 2013	*Tamasheq, Arabic*

This was not the first armed uprising. The Tuareg have resisted foreign rule for a long time.

During colonial times, Alla ag Albachir refused to obey the French administration. A great chief and local hero, they say that when he approached a well even the trees would move aside. In the 1960s, after the French left, his son led the Tuareg uprising against their inclusion within the state of Mali. His camel-borne warriors would swoop down from Saharan mountains to ambush army patrols, only to disappear once more in a haze of dust.

The second rebellion, rekindled in 1990, flared for six years until a ceremonial burning of weapons in the marketplace at Timbuktu. Ten years after this 'Flame of Peace', former rebels took up arms again, igniting a third rebellion. By this time, the insurgents were clear about what they wanted: an independent Tuareg state.

Traditionally nomadic herders ranging throughout the Sahara, these are not conventional Muslims. Known by outsiders as the Blue People because of their indigo-dyed robes, the Tuareg retain many customs and rites that predate their conversion to Islam. Their society is matrilineal, inheritance and descent following the female line. It is the men who veil their faces in front of women, not the other way around.

Veiled fighters realized their dream of a Saharan homeland in the uprising of 2012, announcing their new state to the world in a modern manner: simultaneously on their website and in a Paris television studio. But the dream was short-lived. Concerns over radical Islamist militias prompted the return of the French army, and Tuaregs to withdraw their independence claim. The 'war on terror' allows Mali once more to suppress northern discontent, but they cannot silence the past. This was not the first Tuareg uprising. It is unlikely to be the last.

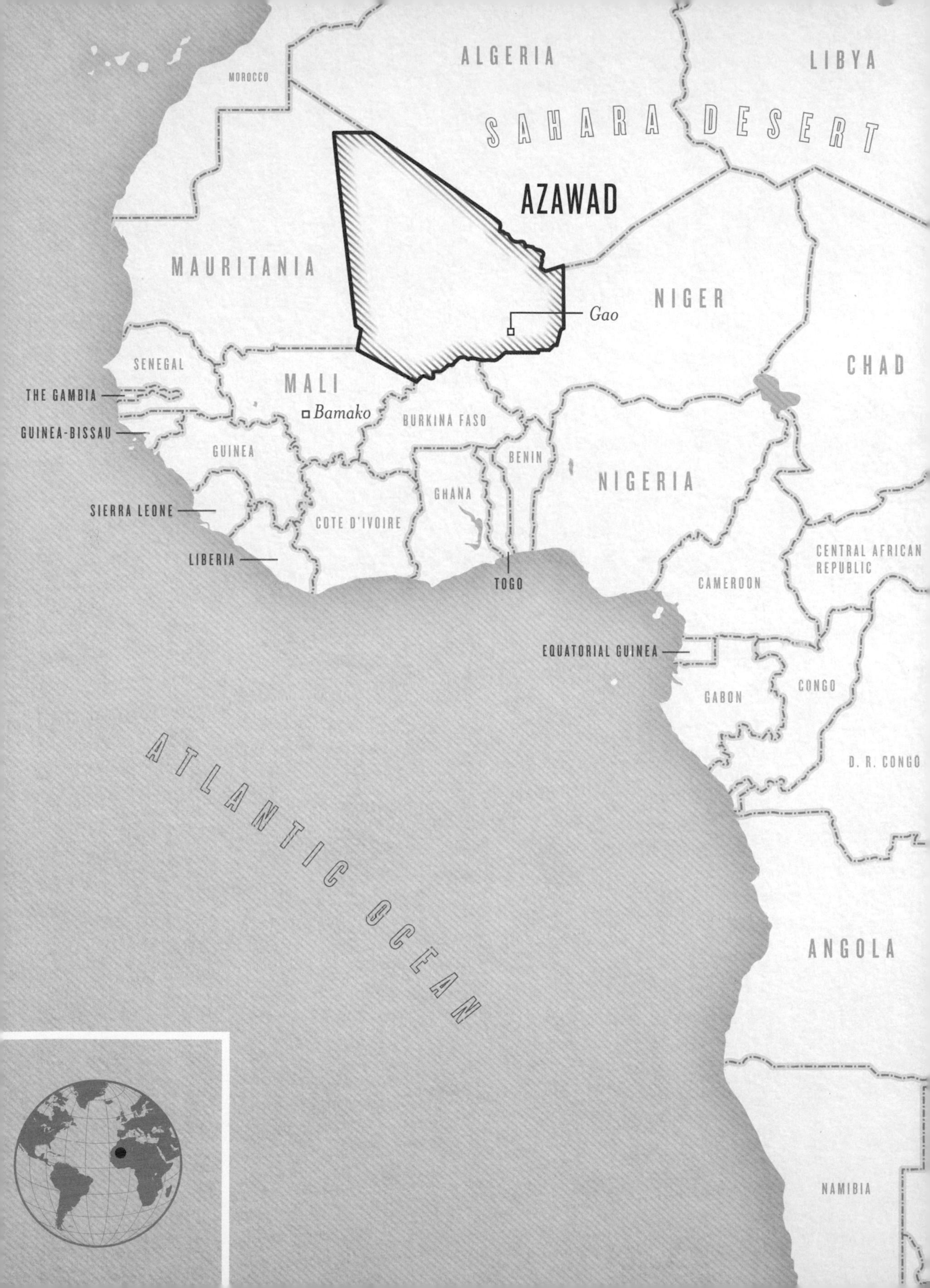

ALGERIA
LIBYA
MOROCCO
SAHARA DESERT
AZAWAD
MAURITANIA
NIGER
Gao
CHAD
SENEGAL
MALI
THE GAMBIA
Bamako
BURKINA FASO
GUINEA-BISSAU
GUINEA
BENIN
NIGERIA
SIERRA LEONE
GHANA
COTE D'IVOIRE
LIBERIA
TOGO
CENTRAL AFRICAN REPUBLIC
CAMEROON
EQUATORIAL GUINEA
CONGO
GABON
D. R. CONGO
ATLANTIC OCEAN
ANGOLA
NAMIBIA

CABINDA

Former Portuguese protectorate recognized as the thirty-ninth African territory to be decolonized by the Organization of African Unity before being annexed by Angola.

DECLARED:

1 August 1975

CAPITAL:

Tchiowa

(Cabinda City)

POPULATION:

688,285

AREA:

7,823 km²

CONTINENT:

Africa

LANGUAGE:

Ibindo, Fiote, Portuguese

Malongo is a gated community. It is a small town, largely inhabited by American ex-pats, with its own electricity, running water, golf course and shopping zones. An on-site greenhouse grows produce for their mess hall salad bar. Egrets strut the manicured lawns and fruit bats roost in the branches of the baobab trees.

Malongo is an enclave within an enclave. It is enclosed within the province of Cabinda, itself swallowed by Angola, although separated from the rest of that country by a strip of the Democratic Republic of Congo. From the beach and high points in Tchiowa or Cabinda city, you can see why Malongo is here. Stretched out across the horizon are the oil platforms and their bright orange flares – where day and night the Cabinda Gulf Oil Company Ltd pumps oil from beneath the Atlantic Ocean.

The foreign residents of Malongo seldom drive beyond the gates of their community. When they want to leave, the Company runs a helicopter service to the international airport. Malongo is surrounded by minefields, helicopter gunship patrols and concentric barbed-wire fences. This is because Malongo is a target for Captain Bonga-Bonga and his men.

The captain is a hero in Cabinda. A soldier for most of his life, in the 1970s he fought for Angola against Portugal, the colonial power. Now he fights for his homeland, Cabinda. The hit-and-run tactics he learned against the Portuguese now serve him against Angola. He is a charismatic leader. He once freed sixty-nine prisoners, including seven of his own guerrillas, from the Cabinda city jail without a shot being fired. The city's jailors knew his grievance was not with them but with the Angolan government. The captain fights for the people of Cabinda, who see few benefits from the oil extracted from beneath their shores by the foreign inhabitants of Malongo.

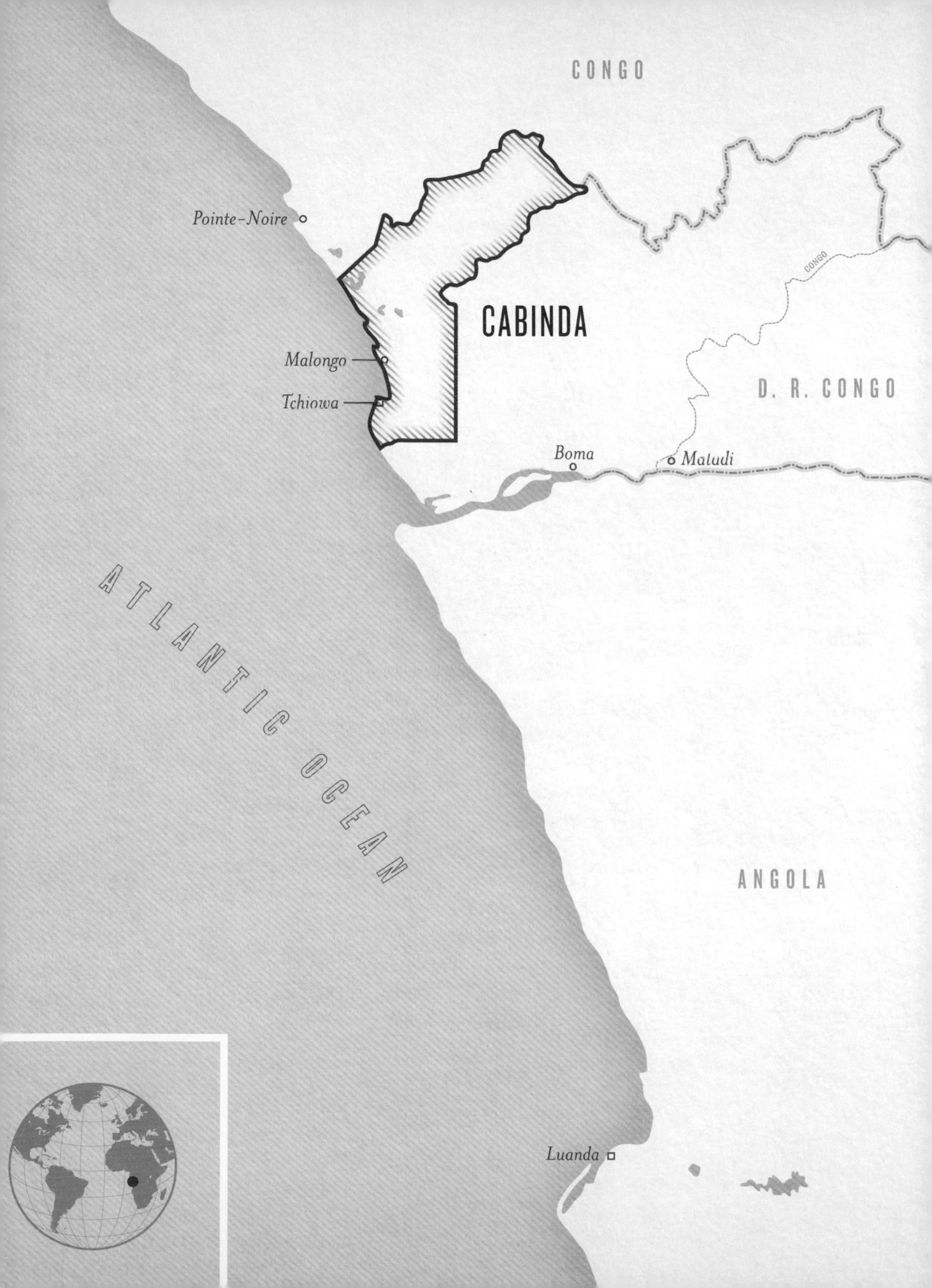

CONGO
Pointe-Noire
CABINDA
Malongo
Tchiowa
D. R. CONGO
CONGO
Boma
Matadi
ATLANTIC OCEAN
ANGOLA
Luanda

OGONILAND

Indigenous Niger delta kingdom.

DECLARED:

2 August 2012

CAPITAL:	POPULATION:
Port Harcourt	*850,000*
AREA:	CONTINENT:
10,000 km²	*Africa*

LANGUAGE:

Khana, Tai, Gokana, Eleme

The turnout for the first Ogoni Day in 1993 indicated something was seriously wrong. Three hundred thousand people celebrated the United Nations Year of Indigenous Peoples by peacefully protesting against the environmental destruction of their homeland. Drums were played and schoolgirls danced in the sunshine. Those without banners held up big white sheets of paper with 'Assassins go home' written across them. People without paper waved leafy branches pulled from the trees. They mopped their brows with handkerchiefs while Ken Saro-Wiwa, a celebrated author and Ogoni spokesman, called for reparations and an end to their mistreatment.

Their protest was against thirty-five years of oil exploitation in their territory, the fertile delta plains of the Niger River where Ogoni people have fished and farmed for centuries. Plains that were now rather less fertile, thanks to spillages, slicks, fires and blowouts. Mr Saro-Wiwa called it an ecological war against ethnic Ogonis and held the Nigerian government and the oil industry responsible. The multinational Shell Petroleum Development Company was declared persona non grata in Ogoniland.

Mr Saro-Wiwa and his colleagues rallied international support for their plight but government troops arrived in response to Ogoni demonstrations. Saro-Wiwa was repeatedly prevented from travelling abroad, harassed and arrested, while the Ogoni suffered escalating violence — much of it, they thought, orchestrated by the military.

In May 1994, four Ogoni leaders were attacked by a mob and killed. The exact sequence of events leading to the murders remains controversial, but many suspected agents provocateurs. Saro-Wiwa and others were arrested, tried and sentenced to death. The UK prime minister at the time called it 'a fraudulent trial, a bad verdict, an unjust sentence ... followed by judicial murder'. Witnesses at the hanging reported Saro-Wiwa's last words: 'Lord take my soul, but the struggle continues.'

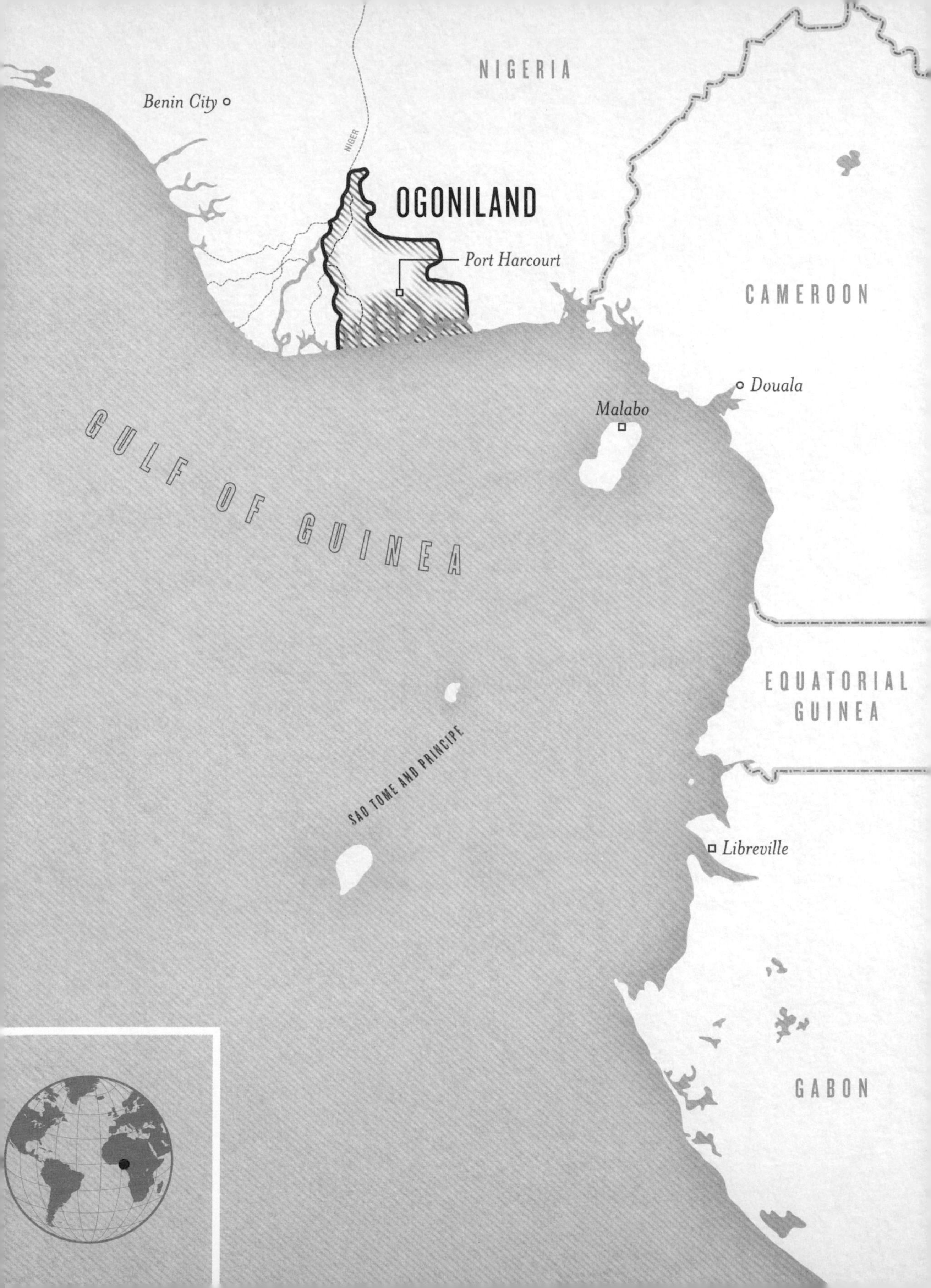

NIGERIA
Benin City
NIGER
OGONILAND
Port Harcourt
CAMEROON
Douala
Malabo
GULF OF GUINEA
EQUATORIAL
GUINEA
SAO TOME AND PRINCIPE
Libreville
GABON

BAROTSELAND

Long-standing monarchy seeking recognition as Africa's newest state.

DECLARED:

8 September 2011; 27 March 2012

CAPITAL:

Mongu

POPULATION:

3,500,000

AREA:

126,386 km²

CONTINENT:

Africa

LANGUAGE:

Silozi, English, 37 other tribal languages

Barotseland is traditionally a mobile kingdom. Every year, as Zambezi River floodwaters seep slowly into their pastures, they up sticks and move to higher ground. This annual migration is celebrated in a ceremony known as Kuomboka, literally 'to get out of the water'. When the moon is full, the thundering of huge drums calls the royal paddlers to assemble from far and wide. Wearing bright red berets, and accompanied by jubilant singing, they propel the royal barges towards the wet-season capital. It is the signal for the king's subjects to load their possessions into dug-out canoes and join the flotilla, leaving their now waterlogged villages behind for another year.

This has been the way of things for as long as anyone can remember. The kingdom has a history stretching back five centuries, but during colonial times Barotseland was a British protectorate, a status allowing greater autonomy than the rest of the area with which it was governed. Northern Rhodesia had colonization; Barotseland had colonization lite. When independence loomed in the early 1960s, the king was persuaded to muck in with what would become the new country of Zambia, on condition that Barotseland maintained that element of self-rule. An agreement was signed allowing the monarchy to pass its own laws over many local matters, including hunting and bushfire control, internal taxation and the beer supply. The deal was called the Barotseland Agreement 1964. BA64 for short.

Only BA64 was never implemented. Successive Zambian governments promised, then failed to honour the deal for the kingdom to enjoy autonomy, systematically ignoring and rebuffing all arguments to the contrary. By 2011, Barotseland's royal household had had enough. A deal is a deal only if honoured by both sides. They pulled out, promising a peaceful disengagement from Zambia, a move denounced in Zambia's capital, Lusaka, as tantamount to treason.

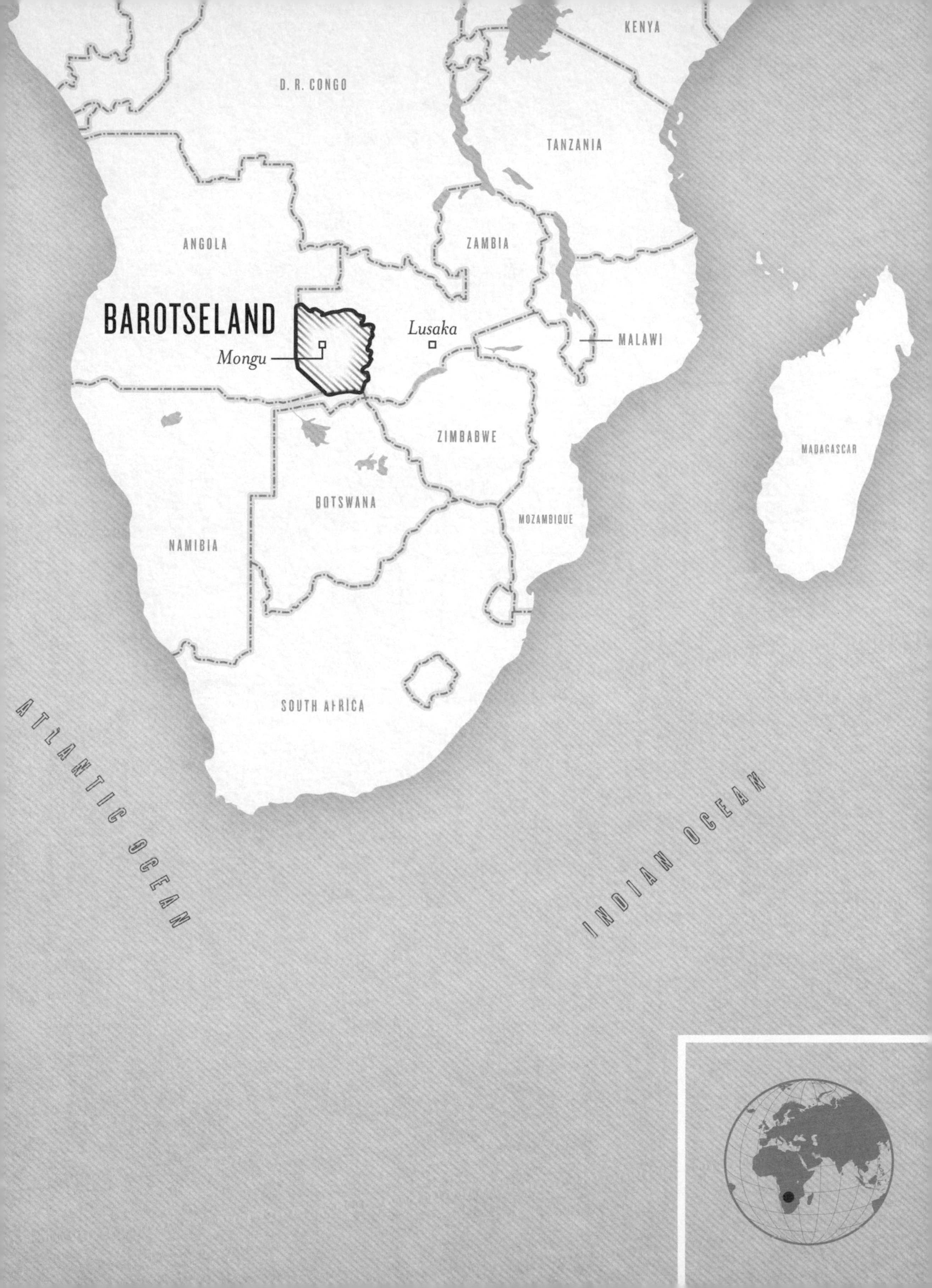

KENYA
D. R. CONGO
TANZANIA
ANGOLA
ZAMBIA
BAROTSELAND
Lusaka
Mongu
MALAWI
MADAGASCAR
ZIMBABWE
BOTSWANA
MOZAMBIQUE
NAMIBIA
SOUTH AFRICA
ATLANTIC OCEAN
INDIAN OCEAN

PONTINHA

Former stronghold of the Knights Templar granted sovereignty in 1903.

DECLARED:

9 October 1903

CAPITAL:

Fort of Saint José

POPULATION:

4

AREA:

178 m^2

CONTINENT:

Africa

LANGUAGE:

Portuguese

They say the steps up to the fort were hacked out of the rock in the early 1400s. Two sea captains, sent by Prince Henry the Navigator, stopped 700 kilometres off the coast of West Africa and made the small island their base for expeditions into nearby Madeira. For hundreds of years this fortress, rearing up out of the rugged basalt, remained a staging post for explorers and merchants sailing into the Atlantic and beyond. During the Napoleonic Wars, British forces used it as a military base and penal complex.

At one time, the fort became a stronghold of the Order of Knights Templar, descendants of the warriors who occupied Jerusalem during the Crusades. The connection sparked an archaeological frenzy when, in 2010, a nail from the time of Christ's crucifixion was found in an ornate casket buried beneath the ancient battlements, alongside three skeletons and three swords. One of the swords sported the Knights Templar cross engraved upon its blade.

The fort remains, but Pontinha is a tiny island no more. Much of it was blasted away in the nineteenth century when Funchal's harbour wall was built, but what remained was sold in 1903 under a royal charter by King Carlos of Portugal. The regal letter granted sovereignty to the owners of the 'fort and the rock upon which it stands'.

Today, if you stroll along the concrete harbour wall you can't miss the red carpet that covers the first few steps across the Pontinhan border. A sign declares 'VISITORS WELCOME (AT OWN RISK)'. Renato Barros, a schoolteacher, is the proud owner of the fort and wants the rights due to all countries, including the full 200 nautical miles of territorial waters extending out into the Atlantic. Admission is free, but donations are welcome to help with restoration work.

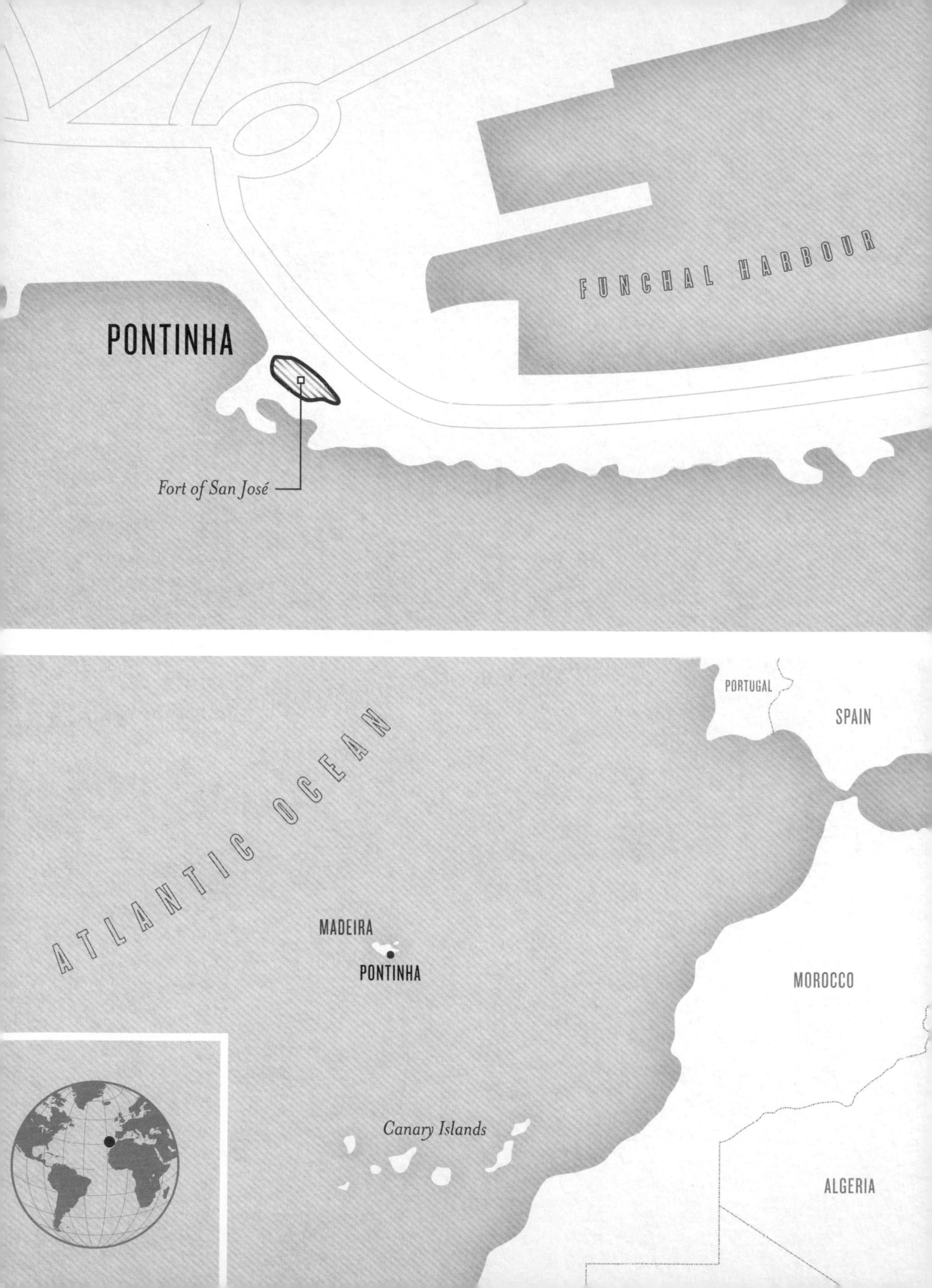

FUNCHAL HARBOUR
PONTINHA
Fort of San José
PORTUGAL
SPAIN
ATLANTIC OCEAN
MADEIRA
PONTINHA
MOROCCO
Canary Islands
ALGERIA

SAHRAWI

Also known as Western Sahara

Africa's last colony.

DECLARED:

27 February 1976

CAPITAL:

El Aaiun

POPULATION:

514,000

AREA:

266,000 km²

CONTINENT:

Africa

LANGUAGE:

Arabic, Spanish

On the outside, they call it the Wall of Shame, a gargantuan earth embankment second only to China's Great Wall in length. Patrolled by sentries and studded with mines, it consolidates the control of their homeland by an occupying force. Safe on the inside, Moroccans plunder their resources: nearly half the world's phosphate reserves – essential to modern farming – oilfields and rich fishing grounds offshore.

The United Nations has been searching for a political solution here in the arid Western Sahara since Spain, the European colonizers, withdrew in 1976. One colonial power was replaced by another and the Sahrawi people rose up to fight for their rights. After fifteen years of fruitless combat they laid down their arms and lobbied for a referendum, allowing Sahrawis the right to vote for independence or permanent integration with Morocco. But the referendum has never taken place.

Decades of endless negotiations have failed to determine who should be entered on the electoral register. Sahrawis demanded the removal of Moroccan settlers; Morocco argued that settlers be allowed to vote, now that they are citizens of the region, whereas many Sahrawis, located in dusty tented camps in neighbouring Algeria, should not be registered since they do not live in the country.

One of the longest decolonization conflicts in history has vanished into an endlessly twisted discussion on procedure. Meanwhile some Sahrawis have spent forty years as refugees in Algeria. They have reached middle age living in a Kafkaesque transition period without end. They know little more than their prison of sand.

They have never set foot on their native soil, but each night they dream of freedom, of an independent Sahrawi republic. A Sahrawi lawyer sums up their faith in a land they are yet to see. He asks: 'Do people change their religion because they don't see God?'

MOROCCO
Canary Islands
ALGERIA
El Aaiun
SAHARA DESERT
MOROCCAN WALL
SAHRAWI
MAURITANIA
Nouakchott
MALI
NIGER
SENEGAL
Niamey
THE GAMBIA
Bamako
GUINEA-BISSAU
BURKINA FASO
GUINEA
BENIN
NIGERIA
GHANA
SIERRA LEONE
CÔTE D'IVOIRE
LIBERIA
TOGO
CAMEROON
GULF OF GUINEA
EQUATORIAL GUINEA
GABON

NORTH A

MERICA

LAKOTAH

Indigenous American people that unilaterally withdrew from all treaties with the USA.

DECLARED:

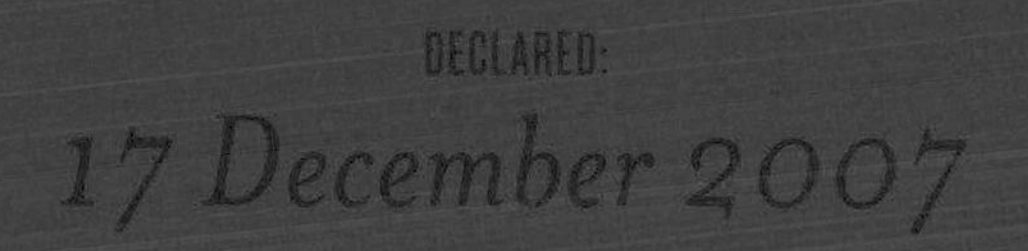

17 December 2007

CAPITAL:

Porcupine

POPULATION:

100,000

AREA:

200,000 km²

CONTINENT:

North America

LANGUAGE:

English, Lakota (Sioux)

They won't take the money. Why should they? To them, the Black Hills are sacred and not for sale. And they were stolen anyway. Accepting the money would legitimize the crime.

In 1868 the Lakota Sioux signed a treaty with the US government that promised the Black Hills would be theirs forever. But just a few years later gold was discovered and the government changed its mind. It reneged on the deal and expropriated the land.

The Lakota name for the Black Hills is Wamaka Ognaka I-cante, 'the heart of everything that is'. According to Lakota creation legend, in the beginning the universe was given a song and a piece of the song is held in each piece of the universe. Except the Black Hills, that is. They hold the entire song. It should come as no surprise to learn that the Lakota have fought for 150 years, on battlefields and in lawcourts alike, for the return of this most spiritual of places.

More than a century after it was expropriated, a US judge awarded compensation for the land – at 1877 prices plus interest. The Lakota are not rich. They languish on a few reservations, ragged scraps of their territory by treaty. By all economic measures these are places of misery and deprivation; not charming, agrarian poverty, but the grimy, squalid, no-hope variety normally associated with urban ghettos. They could use the half-billion dollars of compensation money but still they won't take it. A price tag is only one measure of value, they say.

In December 2007, the Republic of Lakotah was formed. A delegation travelled to Washington DC to deliver their formal withdrawal from the treaties signed with the US government. Not so much a secession as a reassertion of sovereignty. The case of the Black Hills land claim continues.

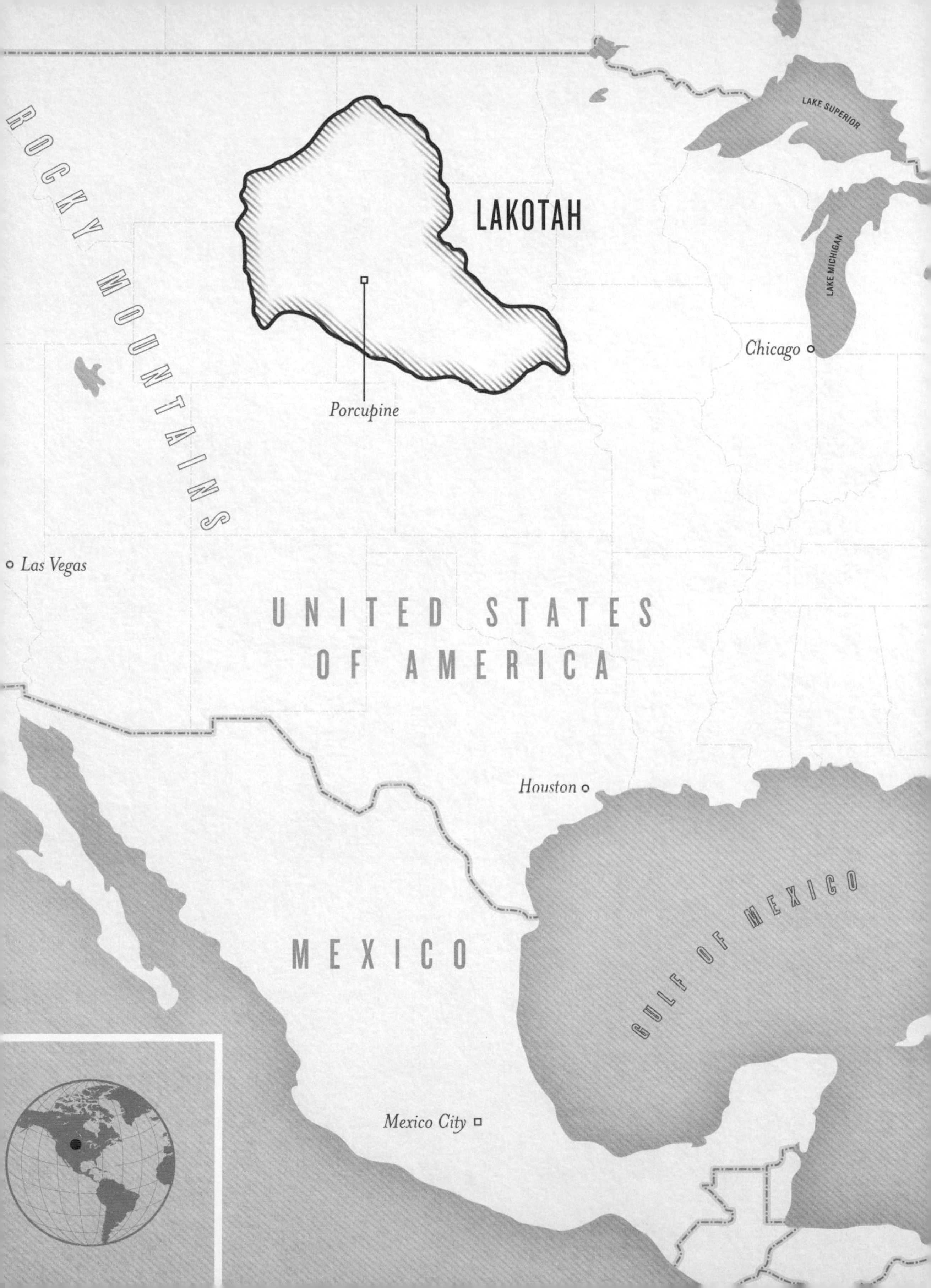
ROCKY MOUNTAINS
LAKOTAH
Porcupine
LAKE SUPERIOR
LAKE MICHIGAN
Chicago
Las Vegas
UNITED STATES
OF AMERICA
Houston
MEXICO
GULF OF MEXICO
Mexico City

REDONDA

Sovereignty declared in 1865 by the first king, whose position is now disputed by several claimants to the throne.

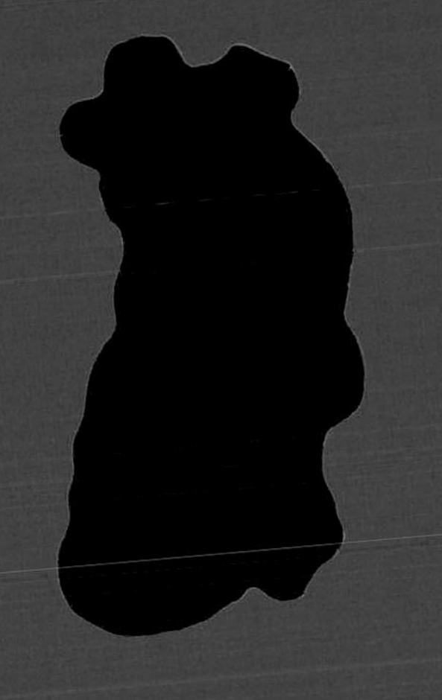

DECLARED:

21 July 1865

POPULATION:

0

AREA:

2 km²

CONTINENT:

North America

DISSOLVED:

1872

FOUNDER:

Matthew Shiell

LANGUAGE:

English, Spanish

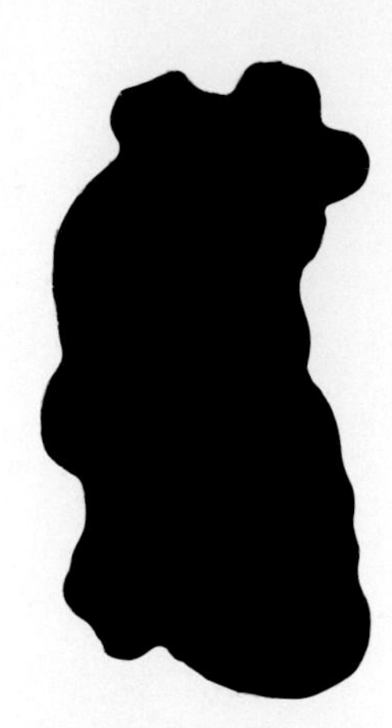

Christopher Columbus himself was the first to record a sighting, on his second voyage of discovery in 1493. Impossibly steep on all sides, only swooping boobies were able to nest on this immense, treeless remnant of an extinct volcano.

Nearly 400 years later Matthew Shiell, a trader from nearby Montserrat, claimed the island as his kingdom. But the short-lived King of Redonda was unable to resist when Britain annexed the land to mine its phosphate-rich guano – the excrement of seabirds accumulated over centuries. A jetty and a cable-hoist were installed and shelters erected. Food was brought in from Montserrat, guano sent out on ocean-going steamers. They shipped the shit to fertilize the fields of Germany and the United States.

Work halted at the outbreak of World War I, and never resumed. A hurricane tore down most of Redonda's buildings, but the craggy kingdom's monarchy persisted in exile. The first king abdicated in favour of his fifteen-year-old son, who left to become a popular novelist in his new home, England. On his deathbed the royal Redondan lineage was bequeathed to a poet, his literary adviser. In the 1940s and '50s several bookish figures of note were given Redondan royal appointments, from Lawrence Durrell to Dylan Thomas, Dorothy Sayers and J. B. Priestley. Falling on hard times, the island was offered for sale to the Swedish royal family, who declined.

In his later years, the poet king spent much of his time in the Alma Tavern in London's Westbourne Grove. It is said that in return for buying the monarch a drink a dukedom or knighthood could be obtained, with a royal inscription on the back of a napkin. And the true current King of Redonda? Unknown. A recent count identified nine pretenders to the throne, testament to the royal confusion perpetrated by the poet.

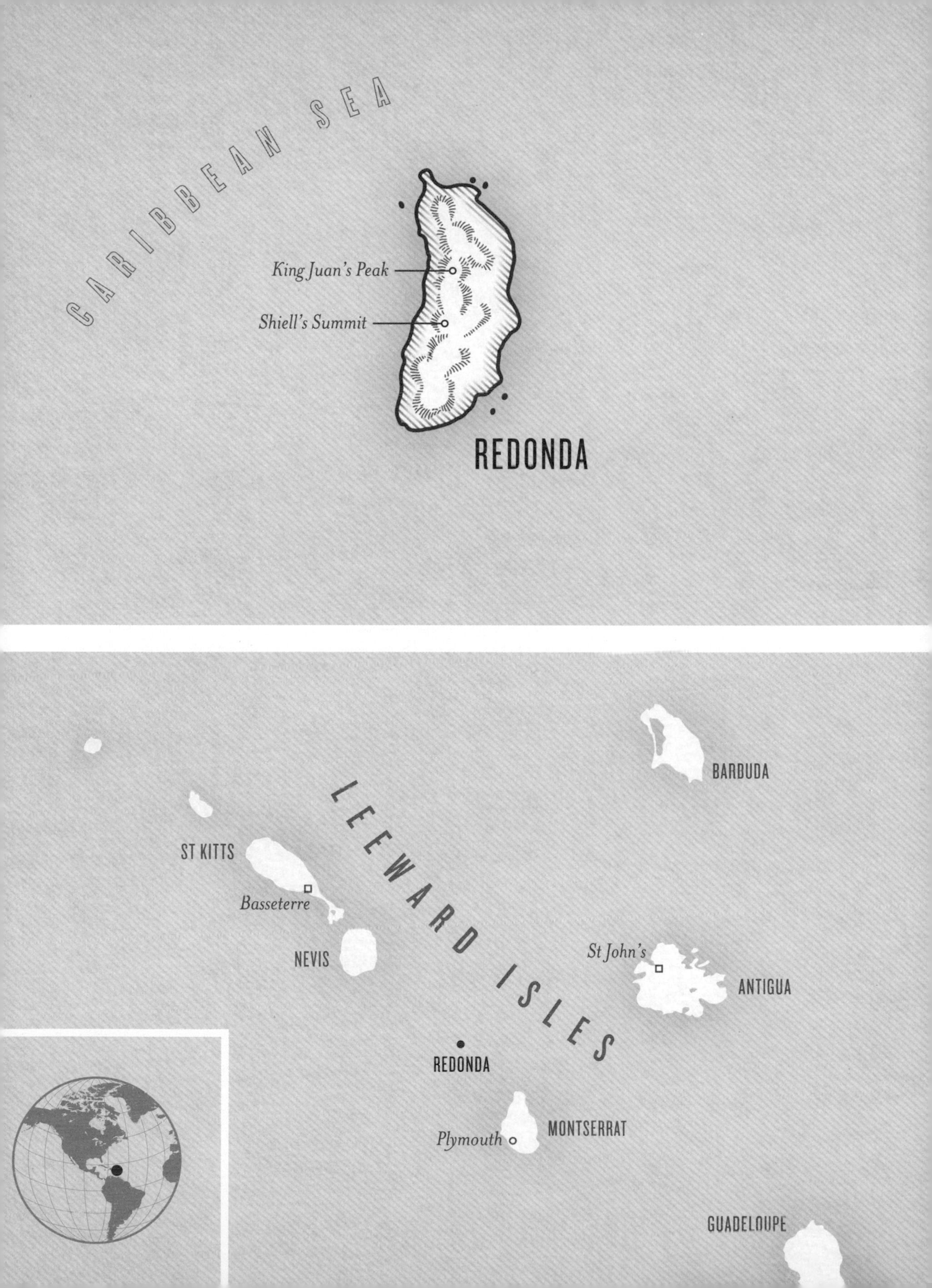
CARIBBEAN SEA
King Juan's Peak
Shiell's Summit
REDONDA
BARBUDA
LEEWARD ISLES
ST KITTS
Basseterre
NEVIS
St John's
ANTIGUA
REDONDA
MONTSERRAT
Plymouth
GUADELOUPE

DINETAH

The largest reservation-based Indian nation in the USA, with a degree of self-government.

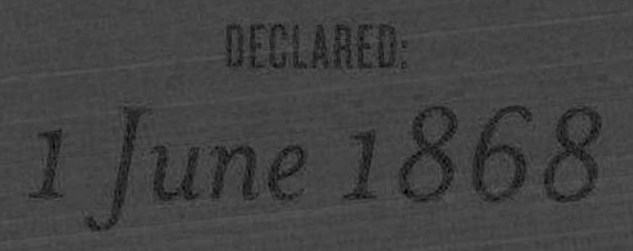

DECLARED:

1 June 1868

CAPITAL:

Tseghahoodzani
(Window Rock)

POPULATION:

300,048

AREA:

71,000 km²

CONTINENT:

North America

LANGUAGE:

Navajo, English, Hopi

They put a stop to it in 2005. When the act was finally signed into law, uranium mining and processing in the Navajo Nation ceased at a stroke. Too many Navajo had died from cancer or other diseases associated with radiation exposure. As one campaigner put it, this legislation had just chopped the legs off the uranium monster.

After World War II, mining provided significant income to the Dine, better known as Navajo, the name given them by neighbouring peoples. Dine veterans, who had used their language to create an unbreakable secret battlefield code while fighting the Japanese, returned home to find work in uranium mines on reservation lands. Unaware of the dangers and without protective clothing, the effects of uranium poisoning ravaged an entire Navajo generation. That was before the worst nuclear accident in US history, which saw 100 million gallons of radioactive mining wastewater released into a local river.

Traditionally, the Navajo's homeland of Dinetah never had precise boundaries but was marked by four sacred mountains at the cardinal compass points. This territory was lost altogether in the mid-nineteenth century, when the Navajo people were marched, at gunpoint, 350 miles to a squalid internment camp. After four years' imprisonment, in 1868 they were handed a treaty to sign. It confiscated 90 per cent of their homeland and allowed them to walk back to what was left.

Over the years additional plots were returned by presidential order until the Navajo Nation now stands as the largest Indian reservation in the USA. Given US citizenship in the 1920s and permitted to teach their own language in schools in the 1960s, the Navajo were granted self-determination in 1975. Hence the authority to ban uranium. But with unemployment on the reservation at nearly 50 per cent and many people lacking basics like electricity and running water, self-determination comes at a price.

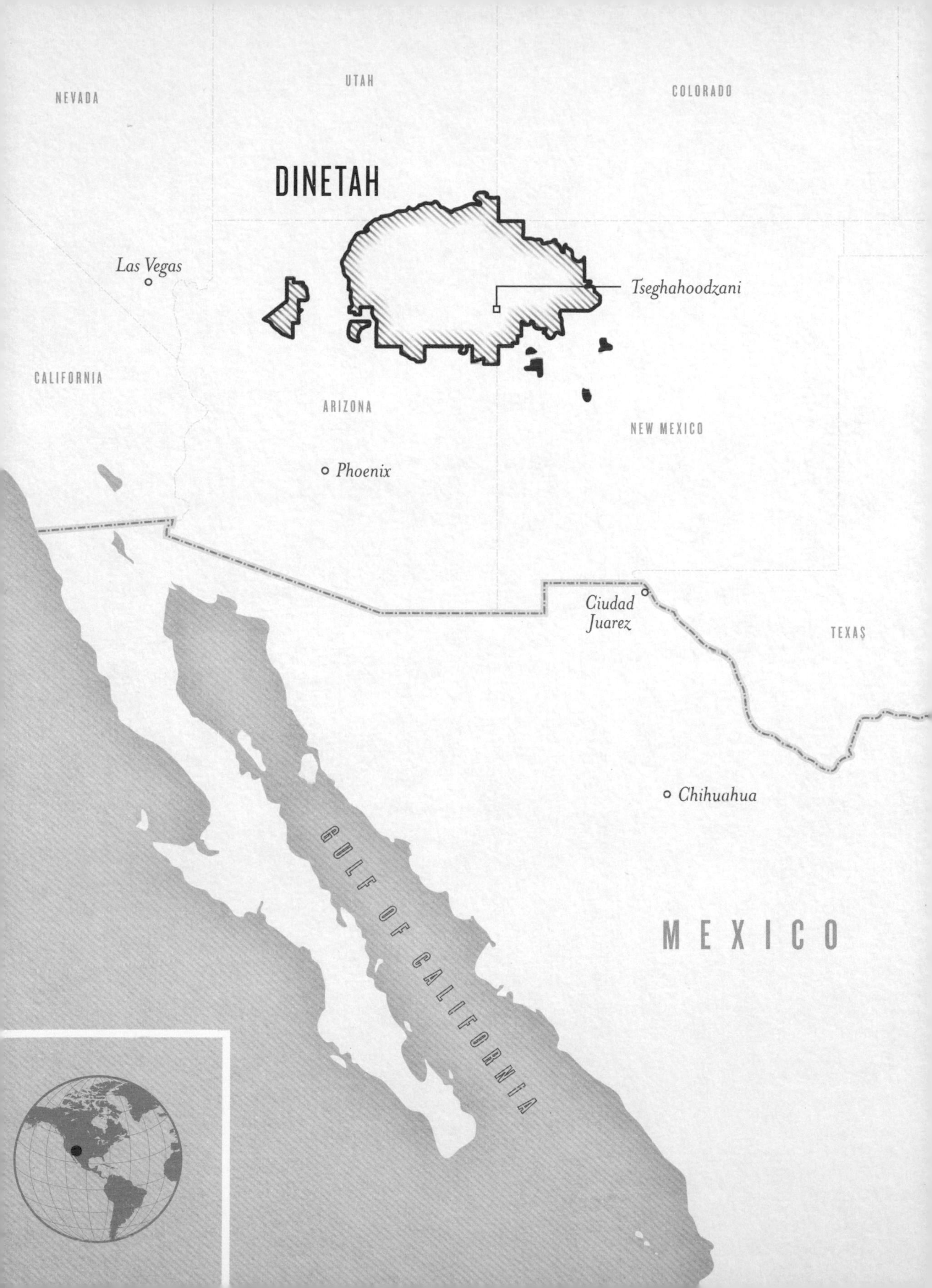
NEVADA
UTAH
COLORADO
DINETAH
Las Vegas
Tseghahoodzani
CALIFORNIA
ARIZONA
NEW MEXICO
Phoenix
Ciudad
Juarez
TEXAS
Chihuahua
GULF OF CALIFORNIA
MEXICO

GREENLAND

Also known as Kalaallit Nunaat

Autonomous part of Denmark moving towards full independence.

DECLARED:

21 June 2009

CAPITAL:

Nuuk

(Godthab)

POPULATION:

57,000

AREA:

2,200,000 km²

CONTINENT:

North America

LANGUAGE:

Greenlandic, Danish

The Queen of Denmark donned national dress for the occasion: knee-high sealskin boots, hoodless anorak and a wide collar decorated with coloured glass beads. She was in Nuuk officially to deliver the law granting self-rule to Greenland's parliament. After the handover ceremony, down by the jetty, a brass band played as men in kayaks made a paddle-past.

The self-governance law had been approved by referendum in the depths of the previous winter. Under its provisions, Queen Margrethe II remained as Greenland's head of state but the world's largest island gained control of its police force and justice system. Greenlandic, spoken by nearly all the island's residents, became their official language.

The Inuit have lived here for 4,000 years, but never ventured far inland. Little point: it's just one huge mass of ice, kilometres thick and devoid of life. There were always plenty of seal, walrus and whale to hunt along the coast. The sparse human population had no centralized administration before the modern colonization of Greenland began in the 1720s. Without any organization above household communities no locals were interested in defending their power, so colonizing Greenland was a peaceful affair. Greenlanders suffered the standard cultural imperialism that comes with colonial status, but never oppression by force. Relations with Denmark were largely cordial. In 1979, Greenland got home rule and in 2009 the Queen turned up to give them self-rule, so ending 300 years under Danish authority. The penultimate step, they say, towards full independence.

To mark the event, the government harpooned a couple of whales, enough to provide celebratory meat for Greenland's entire population. Officials in Nuuk handed it out to local residents in bulging plastic bags; Air Greenland's fleet of small planes was commissioned to distribute the flesh to everybody else: 40,000 people in tiny hamlets dotted round the edges of the great, icy interior.

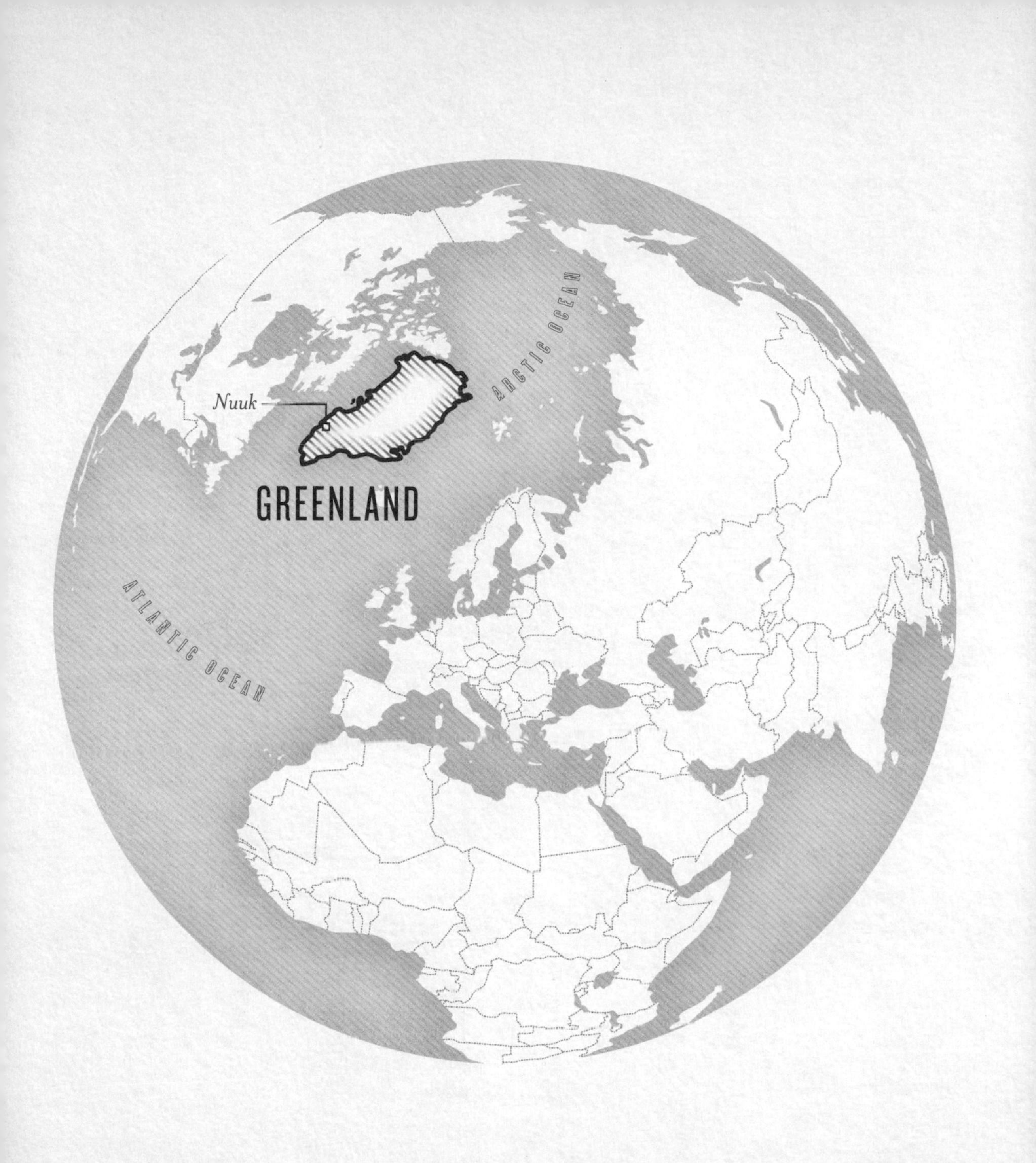
Nuuk
GREENLAND
ARCTIC OCEAN
ATLANTIC OCEAN

LUBICON

Indigenous community never formally ruled by Canada.

DECLARED:

15 October 1988

CAPITAL:

Little Buffalo

POPULATION:

c. 500

AREA:

10,000 km²

CONTINENT:

North America

LANGUAGE:

Cree, English

The children began to complain of headaches and sickness on Friday morning. The principal decided to send them all outside, to get some fresh air, but this changed nothing. The air smelled terrible and the children were still distressed.

Four days passed before the community of Little Buffalo was officially informed of the oil spill. A local government agency sent them a fax saying that 28,000 barrels of crude had leaked from a nearby pipeline. The statement said that no air quality guidelines had been exceeded.

In a sense, the Lubicon Cree of Little Buffalo are victims of a clerical error. At the turn of the twentieth century, the Canadian government dispatched commissioners to sign a treaty with Alberta's indigenous peoples, so allowing the government to govern the province. The treaty-makers simply missed the Lubicon, so they never gave up rights to their territory. In 1939, Canada promised they would create a reserve for the Lubicon, but never got round to doing so. Lubicons just carried on as usual, hunting, trapping and fishing as they always had, since long before the creation of Canada.

Everything changed in the 1970s, with oil and gas development on their land. Thousands of wells were drilled on Lubicon territory and as moose and other game left the area, Lubicons lost their self-sufficiency. Legal redress disappeared in a maze of wrangling, some laws redrafted retroactively to prevent certain claims. In October 1988, Lubicons blockaded their land, asserting an independence they had never given up. Five days later, Lubicon was invaded. Armed police arrived with attack dogs and helicopter support. Twenty-seven people were arrested, though later released.

Dialogue over the land claim remains unresolved, while oil and gas wells keep pumping. Moose and other wildlife have not returned. Chief Ominayak is nervous. As he put it, 'Even if we win, we've lost.'

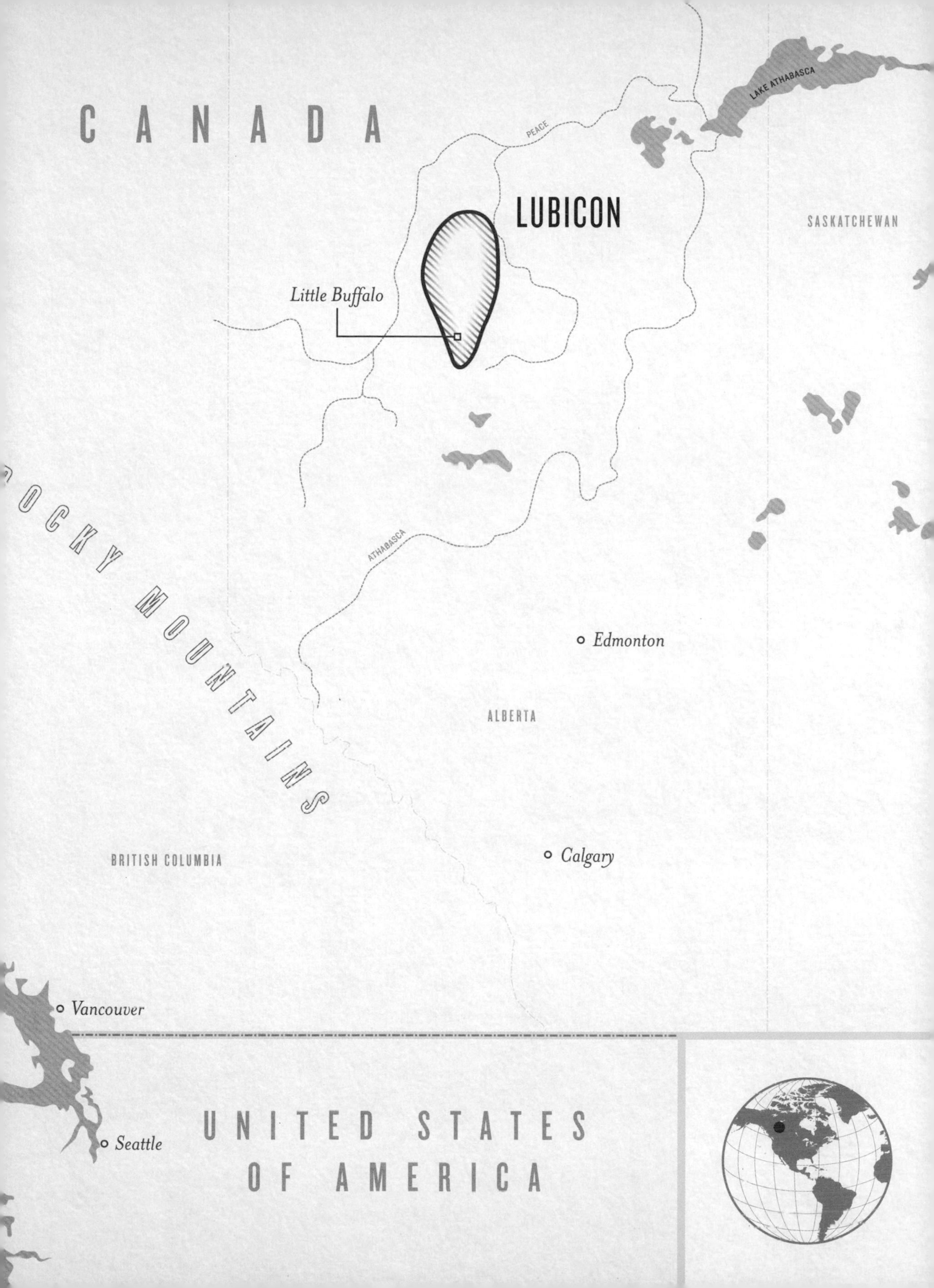
CANADA
LAKE ATHABASCA
PEACE
LUBICON
SASKATCHEWAN
Little Buffalo
ATHABASCA
OCKY MOUNTAINS
Edmonton
ALBERTA
BRITISH COLUMBIA
Calgary
Vancouver
Seattle
UNITED STATES
OF AMERICA

MOSKITIA

Caribbean coastal region seeking a return to self-rule.

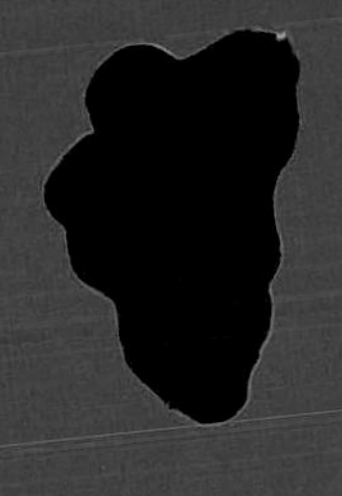

DECLARED:

18 April 2009

CAPITAL:

Puerto Cabezas

POPULATION:

150,000

AREA:

59,600 km²

CONTINENT:

North America

LANGUAGE:

Miskito, English, Spanish, Sumo-Mayangna

The announcement took outsiders by surprise. After electing a new leader on 18 April 2009, the Council of Elders made a declaration: from this day forward, Moskitia would be independent from the rest of Nicaragua. They added that their coastal homeland would be defended, if necessary, by a new indigenous army of approximately 200 men. All companies operating in the region were instructed to desist from paying central government taxes and to pay the Moskitia authorities instead.

Relations between the indigenous Miskito people and Nicaragua's government have seldom been easy. Since their land was annexed by Nicaragua in the late nineteenth century, the previously autonomous Miskitos have received little assistance while facing a litany of woes. These include plagues of rats, periodically devastating hurricanes, and a curious form of contagious hysteria known as GRISI SIKNIS *(crazy sickness). Meanwhile, the government happily allowed generations of outsiders to exploit the Miskitos' jungle territory with impunity. Few of the resultant benefits have been returned to Moskitia, still the poorest and least developed region of Central America.*

Although the general discontent of the Miskito and other indigenous peoples is widely acknowledged, this didn't stop the conspiracy theorists. Some suspected Colombian drug cartels of encouraging separatist sentiments, looking to improve conditions for cocaine trafficking through the region. Others saw the dark hand of US neo-imperialism, seeking to undermine the central government as they did during the Cold War era. But there was no need to invoke sinister subversive forces to explain the breakaway movement. Local feeling was well expressed by Moskitia's new leader, a religious man named Hector Williams, just elected Great Judge of the Nation of Moskitia. Stroking his thin moustache thoughtfully, Mr Williams made reference to 1894, the year his homeland lost its autonomy. 'People have been waiting and waiting for this for 115 years,' he explained. 'Everything has its moment.'

MEXICO
CUBA
HAITI
JAMAICA
DOMINICAN REPUBLIC
BELIZE
GUATEMALA
HONDURAS
Puerto Cabezas
EL SALVADOR
Managua
MOSKITIA
NICARAGUA
COSTA RICA
PANAMA
VENEZUELA
PACIFIC OCEAN
COLOMBIA
ECUADOR
PERU
BRAZIL

SOUTH A

MERICA

MAPUCHE

Indigenous American people seeking a return to self-determination.

DECLARED:

6 January 1641

CAPITAL:

Temuco

POPULATION:

1,700,000

AREA:

650,000 km²

CONTINENT:

South America

LANGUAGE:

Mapudungun, Spanish

It is all about territory. In their own language, Mapuche means 'people of the land'. Trouble is, the Mapuche have spent a long time being dispossessed of their traditional heartland.

The decline has been steady over the last century or so, in harsh contrast to the previous 10,000 years of self-rule. Tucked away in the southernmost tip of the continent, the Mapuche were never conquered by the Incas. Or indeed the Spanish. In fact, they were the first and only indigenous people in South America to have their sovereignty recognized by the Spanish empire. In 1641, the two sides signed a treaty setting the Bio Bio River as the boundary between them.

The river frontier held for another two centuries and more, until in 1881 the Chilean army defeated the Mapuche, seventy years after achieving independence from Spain. At about the same time, on the eastern side of the Andes, Argentina conducted a similar campaign of pacification. All over Mapuche territory the results were the same: their land was handed over to immigrant settlers.

For a century they were killed or evicted, sidelined and suppressed. Forced into overcrowded reservations, many left to find work in cities. Today more Mapuche live in urban areas than in traditional rural communities. But now, perhaps, the tide is turning. In both Chile and Argentina, governments have recognized indigenous peoples' rights, including the right to land. Returning territory is slow, cumbersome and without guarantees. Most Mapuche hold no legal title to areas inhabited by their ancestors, some of which are now in the hands of rich corporations. Whether or not reclaiming their land is the first step back towards self-determination, as many hope, the land itself is vital. This is how Mapuche people define themselves. As one activist put it, 'Without our land, we are not a people.'

BOLIVIA
BRAZIL
CHILE
ARGENTINA
Santiago
Buenos Aires
Temuco
MAPUCHE
ATLANTIC OCEAN
PACIFIC OCEAN
BRAZIL
CHILE
ARGENTINA
MAPUCHE

RAPA NUI

A special territory of Chile, annexed in 1888 under dubious circumstances.

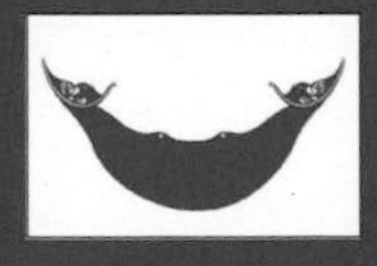

DECLARED:

c. 1200

CAPITAL:	POPULATION:
Hanga Roa	*5,761*
AREA:	CONTINENT:
164 km²	*South America*
DISSOLVED:	LANGUAGE:
1888	*Rapanui, Spanish*

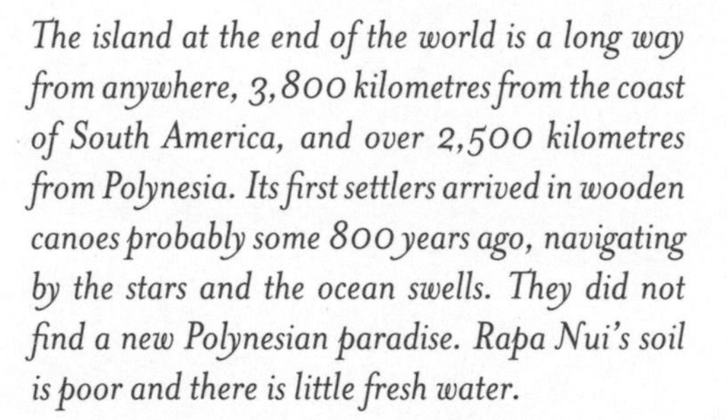

The island at the end of the world is a long way from anywhere, 3,800 kilometres from the coast of South America, and over 2,500 kilometres from Polynesia. Its first settlers arrived in wooden canoes probably some 800 years ago, navigating by the stars and the ocean swells. They did not find a new Polynesian paradise. Rapa Nui's soil is poor and there is little fresh water.

What happened next is a matter for conjecture. What we do know is that several hundred years passed before the first European, a Dutchman, set foot on the windswept Pacific isle. It being Easter Sunday, 1722, he duly named the new Dutch territory Easter Island.

The name has become synonymous with mystery and intrigue. By the time the Dutch arrived, most of the island's giant primeval palm trees had vanished, replaced by enigmatic statues, enormous basalt monoliths embodying the spirits of powerful ancestors. But riddles remain and archaeologists disagree over exactly what went wrong on Easter Island.

The Chilean government took possession of the island in 1888. A naval officer negotiated EL ACUERDO DE VOLUNTADES *('The Voluntary Agreement') with the King of the Rapa Nui and his chiefs, a document in two versions: Spanish, and Rapanui mixed with Tahitian, a spoken language. These versions are not the same. In the Spanish text the chiefs cede sovereignty over the island to Chile; in the Rapanui version Chile offers to be a 'friend of the island'.*

The king and his chiefs could not write, so they signed with an X. During the treaty ceremony, King Tekena grabbed a handful of turf, passed the grass to Chile's naval officer and held on to the soil. To him and his people, this action was worth more than the treaty itself, but to the Chilean government it was the other way around.

RAPA NUI
Ranu Aroi
Hanga Roa
Rano Kau
PACIFIC OCEAN
CHILE
RAPA NUI

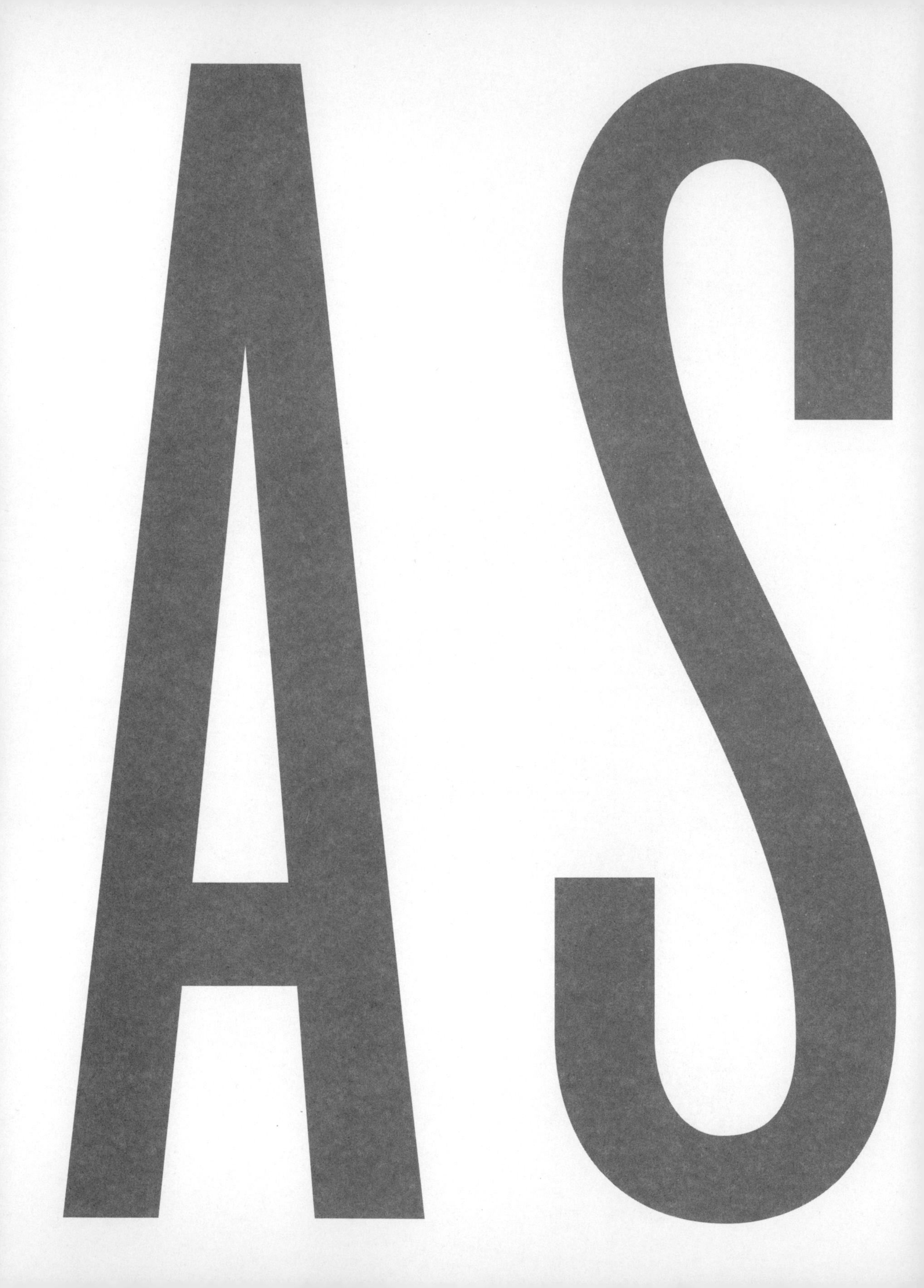
AS

IA

AHWAZ

Also known as Arabistan, Khuzestan

Arabic-speaking corner of Iran seeking a return to self-rule.

DECLARED:

~

CAPITAL:
Ahwaz City

POPULATION:
4,500,000

AREA:
63.238 km²

CONTINENT:
Asia

LANGUAGE:
Arabic, Lurish

They called it the Day of Rage. While the world watched multiple protests escalate into the Arab Spring, local activists using Facebook organized their own anti-government demonstrations. On 15 April 2011, they took to the streets of towns and cities across the region, to mark the anniversary of large-scale unrest six years earlier. Many sources say live ammunition was used against the protesters but few details seeped through the news blackout.

Ahwaz is an Arab region in a non-Arab country. The violence of 2005 had been triggered by reports of an Iranian government policy to forcibly relocate members of the Arab community to other provinces. The government insisted the document was forged, but in Ahwaz they saw just another facet of the foreign occupation of their homeland. Forced displacement, forced assimilation, confiscation of land: they all add up to the same cultural repression.

It is the paradox of plenty. Beneath the sandy soils of this province lie some of the richest oil reserves the world has ever known. Above ground, the Ahwazi Arabs endure extreme levels of poverty, unemployment and illiteracy. It was the discovery of oil in the early twentieth century that sealed the region's fate. The Ahwazis enjoyed self-rule as an emirate called Arabistan until 1925, when Iran's Shah Reza Pahlavi overthrew their ruler and seized total control of the province. Teaching in the Arabic language was banned, traditional Arab clothing outlawed and the name of the region changed to Khuzestan. And so began a long-running, low-level conflict between the government and Arab separatists that every so often escalates into protests with a violent end. Iranian authorities question the spontaneity of such gatherings, seeing foreign influences at work. Some Ahwazi Arabs in exile want a separate state, others seek regional autonomy in a federal Iran. All yearn for an end to the discrimination and deprivation.

IRAN
AHWAZ
Baghdad
Esfahan
IRAQ
Ahwaz City
Basrah
Shiraz
KUWAIT
Kuwait
PERSIAN GULF
BAHRAIN
Doha
QATAR
Riyadh
Abu Dhabi
UAE
SAUDI ARABIA

AKHZIVLAND

Proclaimed independence from Israel in 1971.

DECLARED:

11 April 1971

POPULATION:

2

AREA:	CONTINENT:
0.01 km²	*Asia*

FOUNDER:	LANGUAGE:
Eli Avivi	*Hebrew, Arabic, English*

0 0.25 0.5 0.75 1 KILOMETRES

The azure Mediterranean laps quietly at a sandy cove. Palm trees sway in a gentle breeze. Eli Avivi narrows his eyes at the sun-glint on the water. Balding and barefoot, he is ready for another day in court.

The history of this one-man state and its founder is punctuated by such legal encounters. As a boy in the 1930s and 40s he appeared before a judge many times, charged with sabotage against the British Mandate in Palestine, then the occupying force in the Holy Land. He and his friends were a junior part of the struggle to force Britain out, to make way for the establishment of a Jewish state.

Among the ruins of an old fishing village that is mentioned in the Bible, Avivi illegally constructed a number of huts. In the 1950s they became a peaceful haven for nonconformists from around the world. Dressed in a flowing robe festooned with flowers, Avivi pronounced a teenage couple man and wife a month after they met in Akhzivland. The wedding reception was held on the beach and after the sun slipped below the horizon, guests danced naked around a bonfire. Without permission to conduct marriages from the spiritual authorities in Israel he was liable to legal action, but he got away with a small fine.

Some of Avivi's huts were subsequently torn down by the National Parks authority in 1971, Eli Avivi confronting bulldozers alone with his licensed submachine gun. He was charged with 'establishing a country without permission' but the case was thrown out. Avivi declared that he loved Israel, but he was fundamentally against the government. In Akhzivland he was both inside Israel, but also in his own country, just a short drive up Coastal Route 4 from Nahariya; a country whose national anthem is the gentle sound of ocean waves.

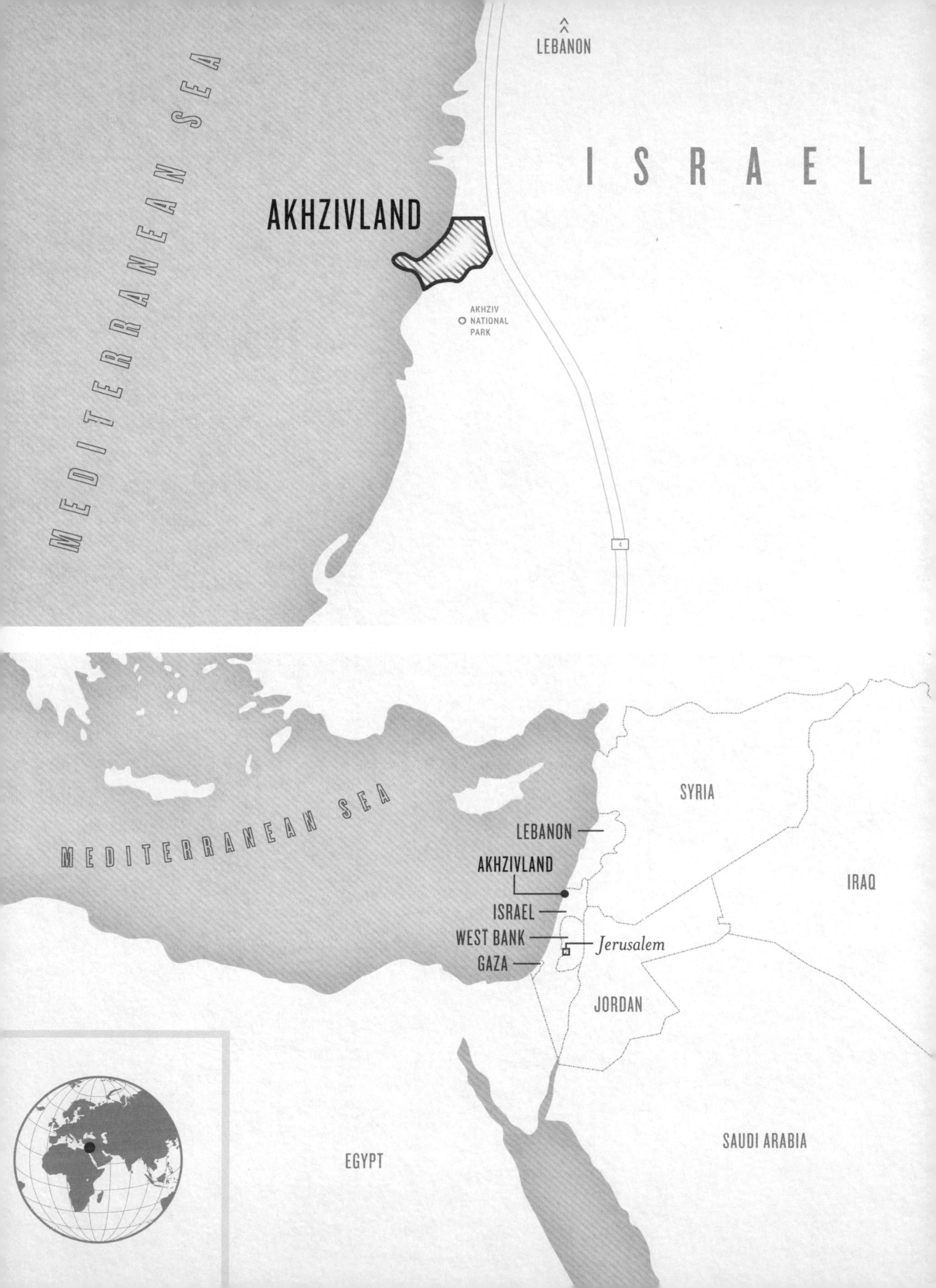

LEBANON
MEDITERRANEAN SEA
ISRAEL
AKHZIVLAND
AKHZIV
NATIONAL
PARK
4
SYRIA
MEDITERRANEAN SEA
LEBANON
AKHZIVLAND
IRAQ
ISRAEL
WEST BANK
Jerusalem
GAZA
JORDAN
SAUDI ARABIA
EGYPT

BALOCHISTAN

Independence declared a day after India and Pakistan, nullified a year later, and declared again after Balochistan's status as a province in Pakistan was abolished.

DECLARED:

15 August 1947; 20 June 1958

CAPITAL:
Kalat

POPULATION:
13,000,000

AREA:
569,800 km²

CONTINENT:
Asia

LANGUAGE:
Urdu, Balochi, Pashto, Brahui

The helicopter gunships came from the east, out of the sun, flying low and in formation. Bristling with rockets and machine guns, they were to prove more than a match for the Marri Baloch people in their goat-hair tents.

Some 15,000 Baloch families had gathered in the fertile Chamalang Valley to graze their flocks of fat-tailed sheep. It is one of the few areas with rich grazing in all of Balochistan, an austere land of rugged desert that respires its own dust. This was the summer of 1974 and most of the menfolk had stayed in the mountains to fight with the guerrillas. Women, children and the elderly had slipped down from the highlands to escape the unremitting bombs and strafing attacks.

Pakistan's military, frustrated by their failure to annihilate the Baloch insurgents, sent in the Huey Cobra gunships to lure the fighters from their mountain hideouts to defend their families. Every Baloch knows the story of Chamalang, a vicious six-day battle with the inevitable, bloody finale.

The pastoral communities of Balochistan were divided in three when Britain drew her imperial boundaries in the nineteenth century. Roughly a third went to Persia, a few were destined for a narrow strip of Afghanistan, and the rest to British India, later Pakistan. The Baloch never asked to be in Pakistan. Forced into nullifying their own independence in 1948, an insurgency arose and is periodically suppressed. In that summer of 1974, the Shah of Iran, fearing that insurrection might spread across the border to the Baloch living in eastern Iran, sent the Huey Cobra gunships with Iranian pilots to help.

The massacre at Chamalang helps to motivate successive insurgencies, each protesting against the continued economic marginalization and political discrimination faced by the Baloch. Each suppression leaves a legacy of hatred to fuel another generation.

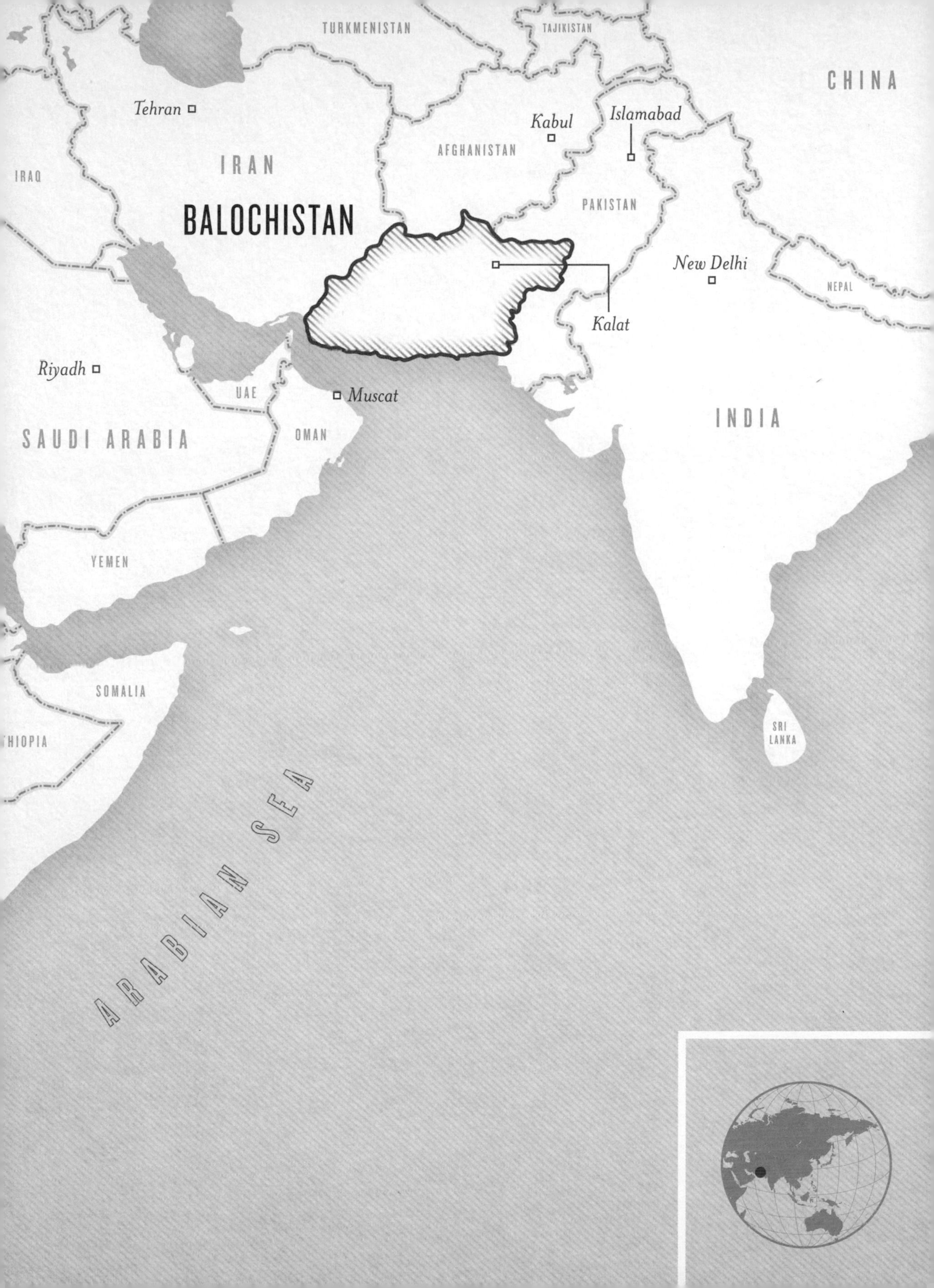
TURKMENISTAN
TAJIKISTAN
CHINA
Tehran
Kabul
Islamabad
AFGHANISTAN
IRAN
IRAQ
BALOCHISTAN
PAKISTAN
New Delhi
NEPAL
Kalat
Riyadh
UAE
Muscat
INDIA
SAUDI ARABIA
OMAN
YEMEN
SOMALIA
THIOPIA
SRI LANKA
ARABIAN SEA

COCOS ISLANDS

Also known as the Keeling Islands

Indian Ocean archipelago ruled by the Clunies-Ross family until 1984.

DECLARED:

7 July 1886

CAPITAL:

West Island

POPULATION:

c. 500

AREA:

14 km²

CONTINENT:

Asia

DISSOLVED:

6 April 1984

FOUNDER:

John Clunies-Ross

LANGUAGE:

Cocos Malay, English

Without the coconuts none of this would have happened. The swaying palms flourished on the atoll and successive descendants of Captain John Clunies-Ross, a Scottish seaman, built a business empire on copra: the dried flesh of coconuts, squeezed for its oil. Labourers from Asia were brought to these isolated ocean specks, nearly 1,000 kilometres from the nearest land, to work the plantations and make Cocos their home.

In 1886, fifty years after Charles Darwin visited on the BEAGLE, Queen Victoria granted all the islands to John's grandson, George, and his heirs in perpetuity. Building of the family home commenced, using bricks and tiles shipped all the way from Scotland. The grand colonial mansion still stands to this day.

What followed is almost another century of personal plantation rule. Workers were given housing, education and medical attention for free, and paid with coloured plastic tokens: Cocos rupees. The currency was minted by the Clunies-Ross family and could be redeemed at the company store, the country's one and only shop. At weekends, families went fishing in the shallow lagoon, launching their boats from the talcum powder-white beach sands.

The last Clunies-Ross to rule the Cocos Islands was known as 'Tuan' (Master). He walked the islands barefoot, with a dagger in his belt. He had complete control and divided opinion. Was he colonial autocrat or benevolent father figure? He ran a working plantation, not a tropical island paradise. Cocos Malays could only earn a living serving the Clunies-Ross family; workers were free to leave the islands, but they could never return.

Under pressure from the modern world, and short of money, Clunies-Ross sold the family sovereignty to the Australian government in 1978. His reign ended six years later when the islanders voted in a UN-backed referendum to integrate fully with Australia. He left the Cocos to live in exile in Perth.

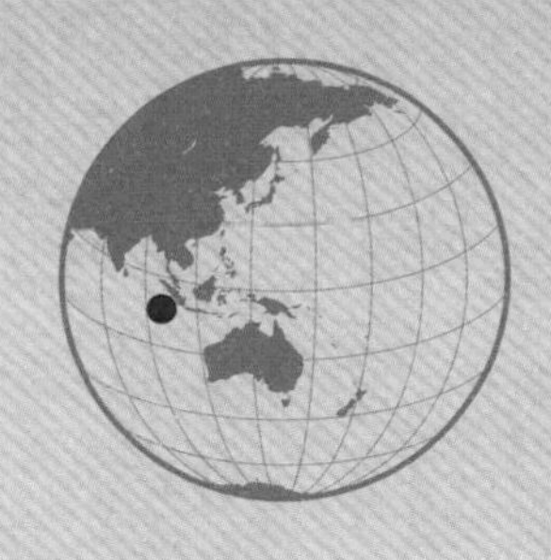

North Keeling Island

NAGALIM

Independence declared a day before India.

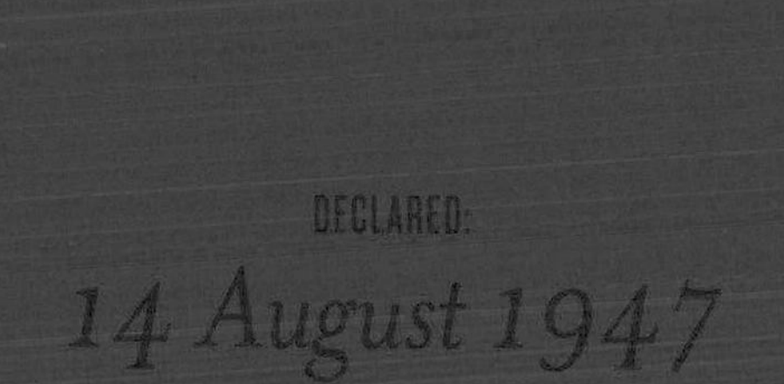

DECLARED:

14 August 1947

CAPITAL:

Kohima

POPULATION:

3,500,000

AREA:

125,000 km²

CONTINENT:

Asia

LANGUAGE:

Nagamese

Rani Gaidinliu was no ordinary teenager. A guerrilla leader fighting British rule, she was sixteen years old and had a price on her head. Hiding out in the Naga Hills with her band of freedom fighters, she taunted the colonials from sepia villages in the clouds. Gaidinliu, it was said, was blessed with magical powers. She could turn bullets into water, come and go at will, like a ghost in the morning mist. When finally arrested by troops of the Assam Rifles, she was convicted of murder and sentenced to life imprisonment. Gaidinliu was freed only after India had become independent. She was thirty-two and had spent nearly half her life in jail.

Before the British arrived in Naga in the early nineteenth century, each hilltop settlement was an independent sovereign state. When villages clashed with their neighbours headhunting was a central feature of each conflict. But the British discouraged headhunting, introduced Christianity, and inadvertently paved the way for Naga nationalism.

The quest for self-determination, now that Naga villages were united, still drove Rani Gaidinliu on her release from prison. But the newly autonomous Indian state, which had ignored her people's own declaration of independence, saw her potential as a beacon for Naga autonomy. Hence it was another decade before they allowed her to return home.

By then Christian Nagas had taken over the separatist struggle, now against the Indians, and the magical girl from the hills was seen as a menace. Threats were made and in 1960 Rani Gaidinliu disappeared again, taking up residence in a cave high above the precipitous green valleys. In later life, Gaidinliu chose negotiation as the means to achieve liberation for her people. With limited success, she died a neglected figure in 1993. But her people continue their struggle, for the dream of a free sovereign Naga nation.

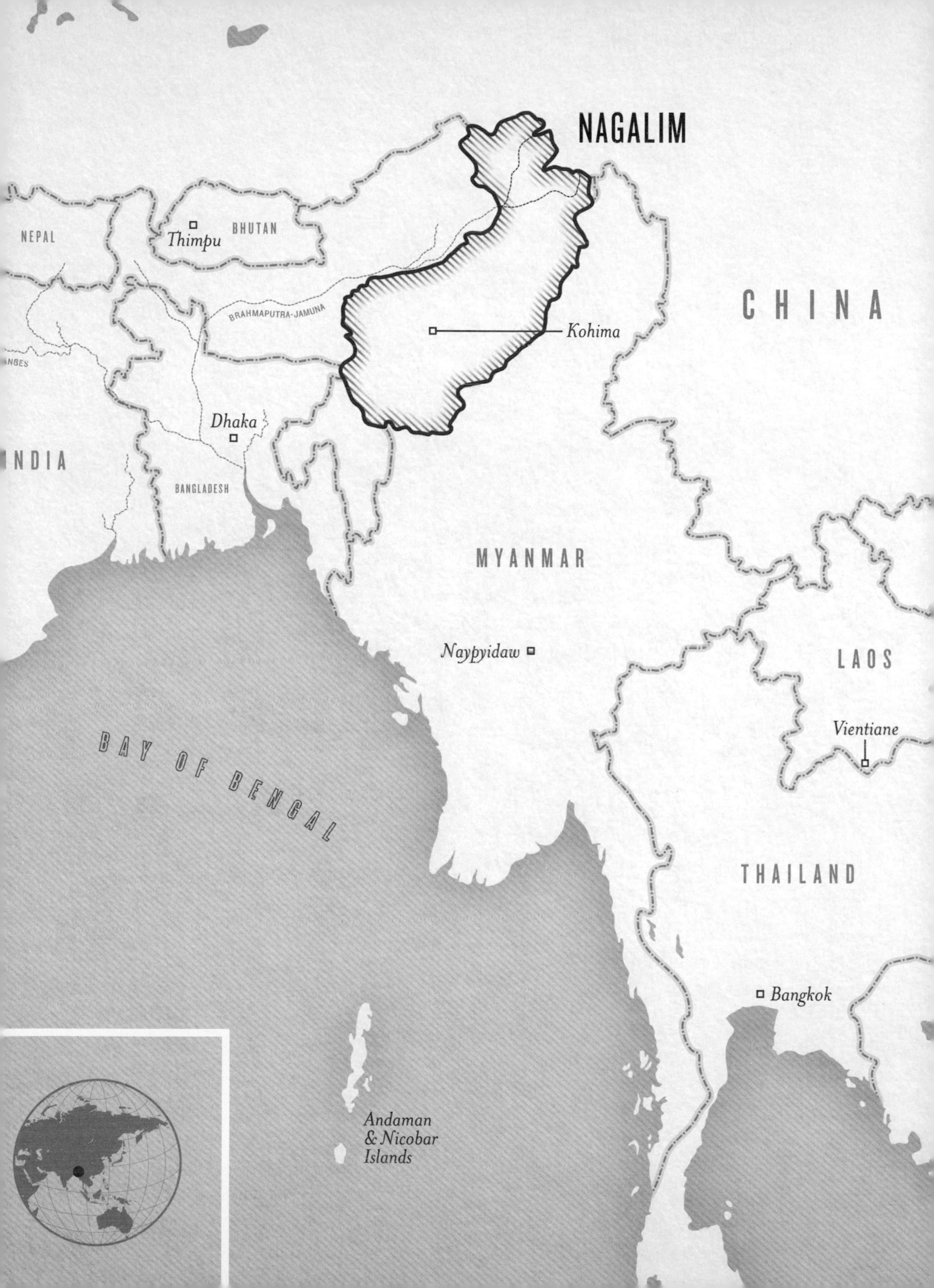
NAGALIM
NEPAL
BHUTAN
Thimpu
BRAHMAPUTRA-JAMUNA
CHINA
Kohima
Dhaka
NDIA
BANGLADESH
MYANMAR
Naypyidaw
LAOS
Vientiane
BAY OF BENGAL
THAILAND
Bangkok
Andaman
& Nicobar
Islands

SIKKIM

Independent Buddhist kingdom annexed by India in 1975.

DECLARED:

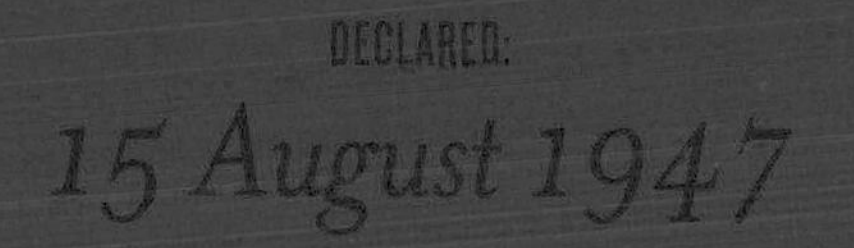

CAPITAL:
Gangtok

POPULATION:
610,577

AREA:
7,096 km²

CONTINENT:
Asia

DISSOLVED:
16 May 1975

LANGUAGE:
Nepali, Sikkimese, Lepcha, Limbu, Newari, Rai, Gurung, Magar, Sherpa, Tamang, Sunwar, English, Hindi

0 25 50 100 150 200 KILOMETRES

It was early in the morning when the last King of Sikkim heard the roar of military vehicles accelerating up the hill towards his palace. Indian troops were soon everywhere, a 5,000-strong force that quickly overwhelmed his 243 palace guards. It was all over before lunchtime, by which time an independent Himalayan monarchy had ceased to exist.

Under house arrest, King Palden Thondup Namgyal was stunned. He had been a staunch supporter of Jawaharlal Nehru who, as India's first prime minister, had accepted Sikkim's continuing monarchy when India became independent in 1947. It was Nehru himself who later said that 'Taking a small country like Sikkim by force would be like shooting a fly with a rifle.' But this was 1975 and Nehru was dead and cremated. His daughter, Indira Gandhi, was in office. Her government had secretly decided to annex Sikkim some years earlier, and commanded their so-called Research and Analysis Wing (RAW) – the secret service – to prepare the ground. RAW quietly stoked pro-Indian sentiments and supported moves against the king. For two years there had been rioting on the streets of Gangtok.

Sikkim has a geopolitical importance that belies its tiny size. Sitting astride two critical passes through the Himalayan mountains, it has great strategic significance for the political manoeuvring between India and China. Indira Gandhi cited such national interest when she made Sikkim the twenty-second state in the Indian union.

Following the Indian army's move into Sikkim, a referendum was organized in which 97.5 per cent voted to join India. Some Sikkimese suggested the result reflected a degree of manipulation by their neighbour. China protested but in 2003 acknowledged Sikkim as part of India. In exchange, India recognized that Tibet was part of China. The agreement was hailed as a breakthrough in India and China's relationship. Ultimately for Sikkim, its geography was its destiny.

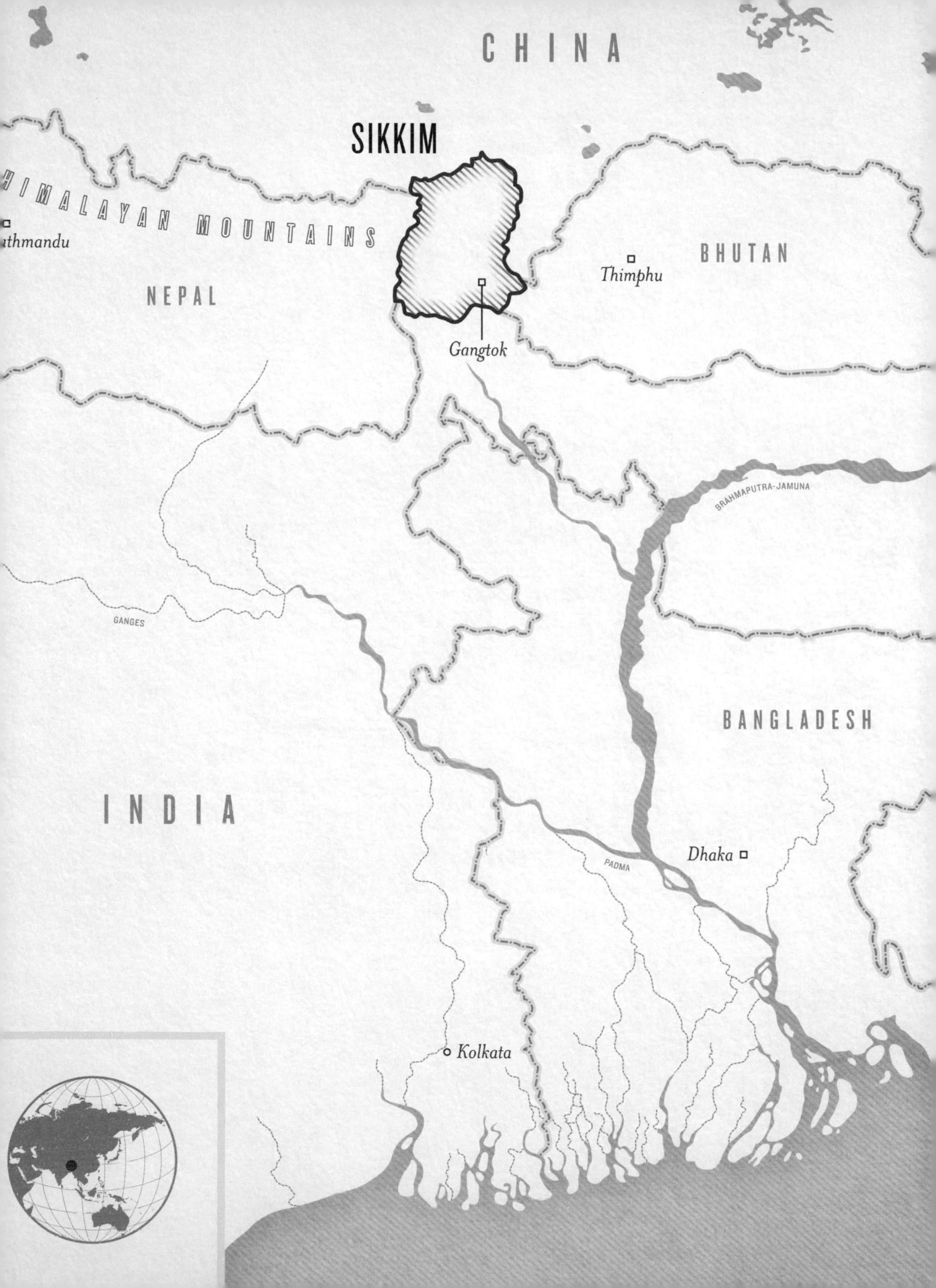
CHINA
SIKKIM
HIMALAYAN MOUNTAINS
thmandu
NEPAL
Thimphu
BHUTAN
Gangtok
BRAHMAPUTRA-JAMUNA
GANGES
BANGLADESH
INDIA
PADMA
Dhaka
Kolkata

TAIWAN

Also known as Republic of China

Chinese province controlled by a republican government since 1949.

DECLARED:

1 January 1912

CAPITAL:

Taipei

POPULATION:

23,400,000

AREA:

36,193 km²

CONTINENT:

Asia

LANGUAGE:

Taiwanese, Mandarin, Hakka, Formosan languages

0 50 100 200 300 KILOMETRES

Taiwan and China agree: they are both part of the same country and the country is China. But how exactly does that work, with two separate governments, one in Beijing and another in Taipei? They call it the 'Two China Problem'. No one seems to have the definitive answer to this one – except that the two sides have agreed to disagree since 1949, when the Chinese Nationalists lost the civil war and fled the mainland. They took up residence on the island that is Taiwan, as the Republic of China (ROC). The ROC was a founding member of the United Nations and most countries maintained relations with this sole representative of China. Until 1971, when everything changed.

The UN General Assembly passed Resolution 2758, expelling the ROC and recognizing the People's Republic of China (PRC) as the only legitimate representative of China to the UN. Subsequently, the ROC (aka Taiwan) has been isolated from the international system, recognized diplomatically by a dwindling number of mostly small states. At the last count, just twenty-two UN member states agreed that Taiwan was a country. One hundred and seventy-two others did not. Nonetheless, many of the latter group maintain unofficial relations with Taiwan, through trade or cultural bureaux that function suspiciously like embassies.

Mainland China continues to claim sovereignty over Taiwan, its twenty-third province, despite Taiwan's own democratic government, its twenty-three million people and its highly industrialized economy. The feeling is mutual. Taiwan's constitution formally claims sovereignty over mainland China. Bizarrely, the fact that Taiwan does not actively pursue its claim over the mainland has created the greatest political paradox of all. If Taiwan officially withdrew its territorial claim over mainland China, to match the real-world situation, many believe the government in Beijing would view it as a declaration of independence, and respond with considerable force.

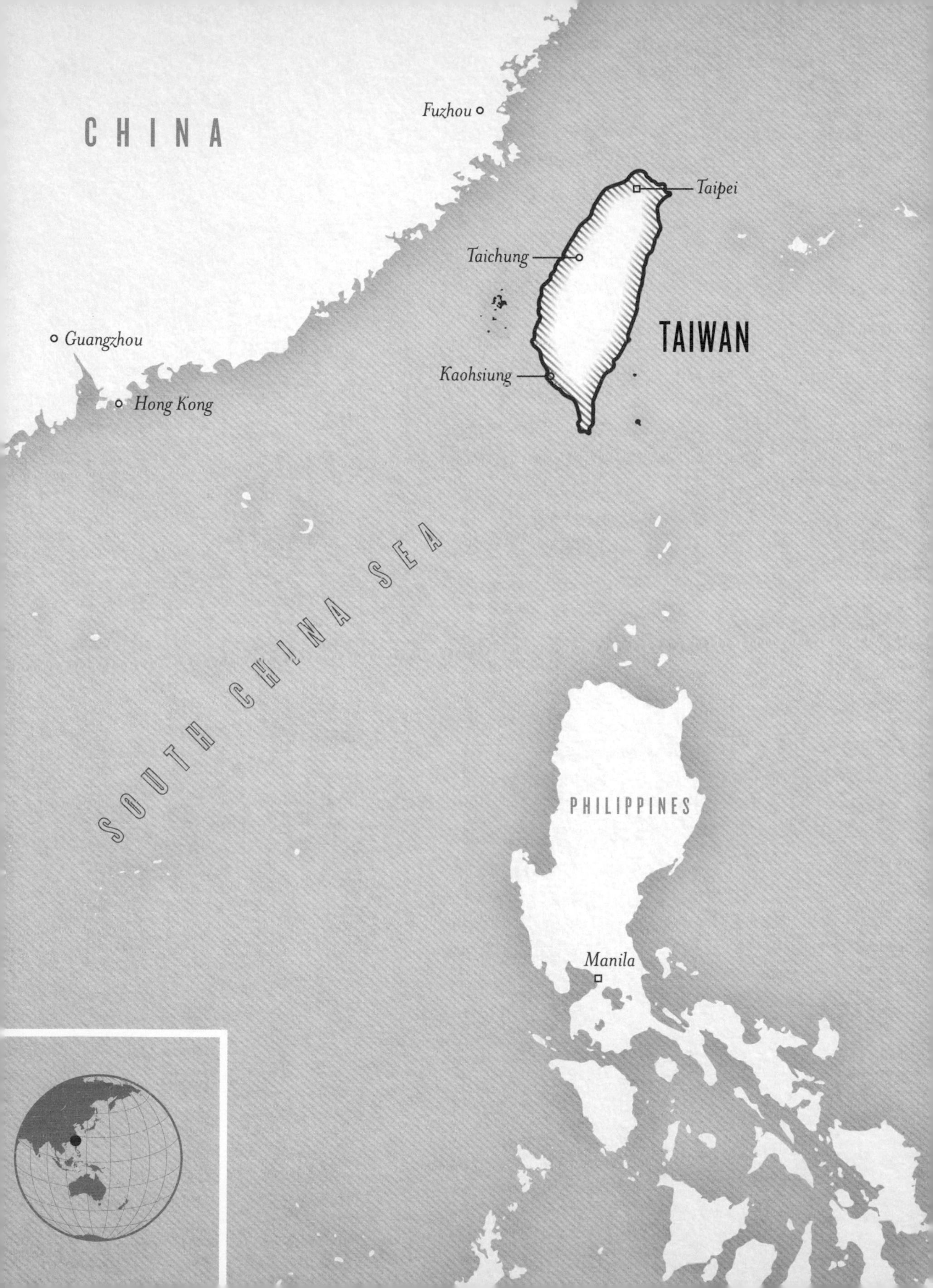
CHINA
Fuzhou
Taipei
Taichung
TAIWAN
Guangzhou
Kaohsiung
Hong Kong
SOUTH CHINA SEA
PHILIPPINES
Manila

TIBET

The spiritual leader of Tibet fled in 1959 after a failed uprising against Chinese rule.

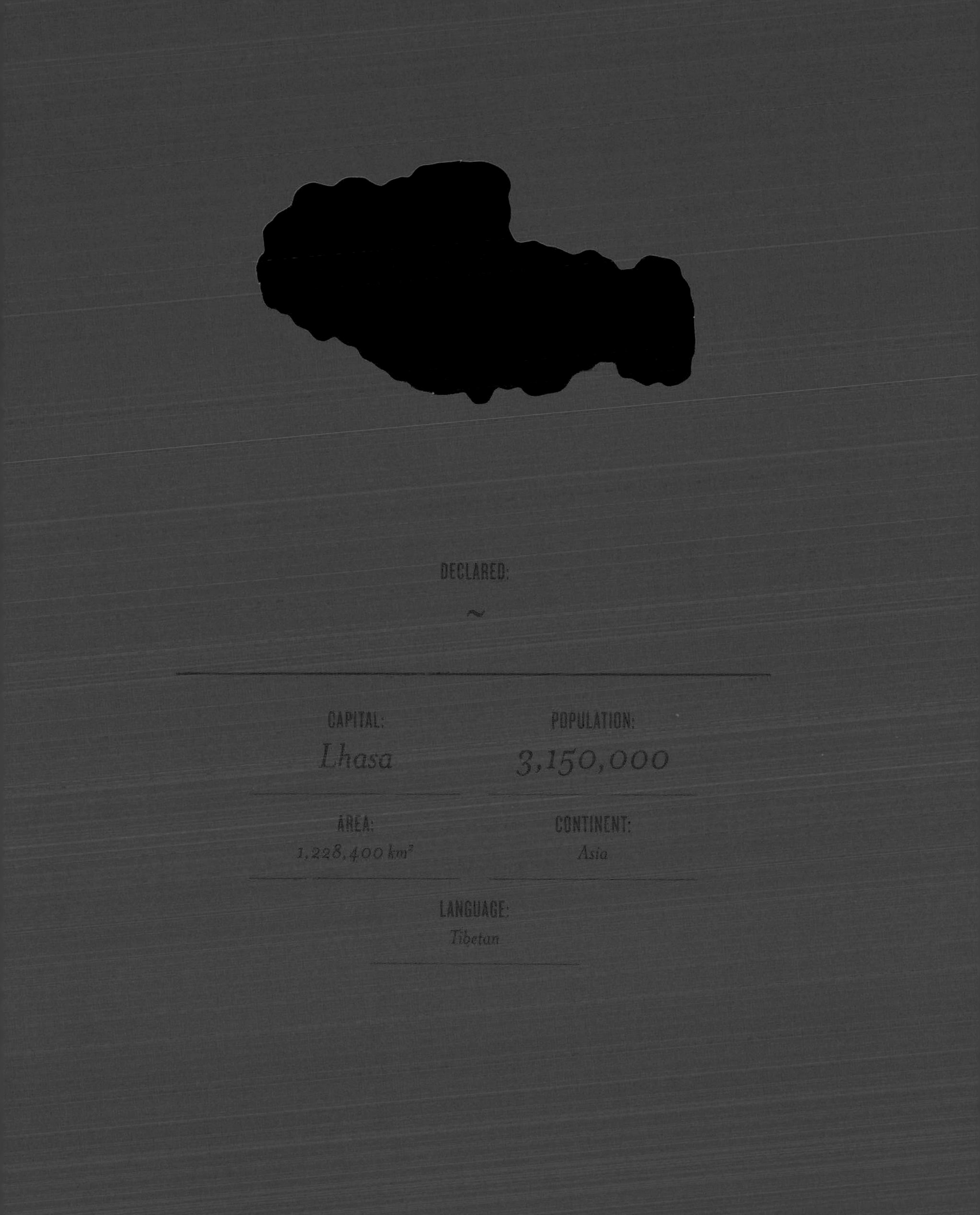

DECLARED:

~

CAPITAL:

Lhasa

POPULATION:

3,150,000

AREA:

1,228,400 km²

CONTINENT:

Asia

LANGUAGE:

Tibetan

0 250 500 750 1000 KILOMETRES

Nobody crossed the mountain pass in winter. The snow was too deep. But this was an emergency, so every man and woman in the village worked through the night to clear the way for his escape route.

It was March, the third Tibetan lunar month, 1959. His Holiness the Fourteenth Dalai Lama had left Lhasa in secret more than a week earlier. As the Chinese People's Liberation Army began firing artillery shells at his palace, he rode away on a horse under cover of darkness. China's official explanation was that he had been kidnapped by bandits.

Following a course prescribed by an oracle, His Holiness arrived in the village of Tsona dressed in a brown knitted cap and wearing dark glasses against the snow-glare. He spent the night in the village monastery, where strings of coloured prayer flags stretched from the compound's mud walls up to the biscuit-coloured cliffs behind. He conducted evening prayers and meditation among the flickering yak butter lamps and old silk wall hangings while outside, high in the pass, the villagers stamped the snow down so the horses could cross.

A half-moon was floating in a perfect blue sky when he departed next morning, the residents of Tsona following his entourage saying prayers as they went. The perpetual parade moved in a constant hum generated by the muttering of mantras. People living along the wayside emerged from their homes to make prostrations as the horses struggled past. The route over the Himalayan mountains was precipitous and rocky, slippery and dangerous.

Over the pass, down through a forest and beyond, the spiritual leader of Tibet was offered asylum in India. The Dalai Lama settled in the city of Dharamsala, he and his followers moulding an area now known as Little Lhasa. And so began more than half a century of exile.

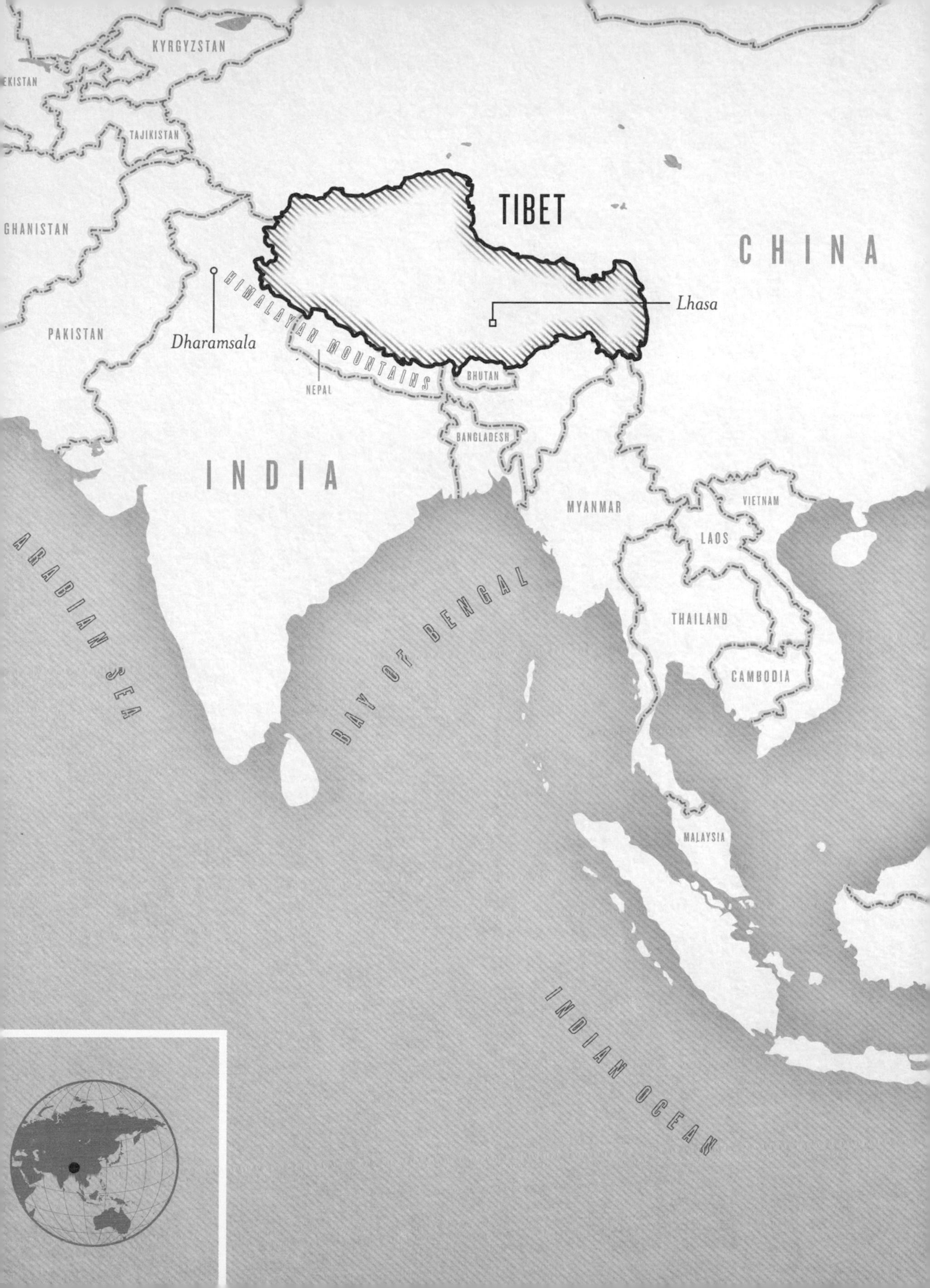

KYRGYZSTAN
EKISTAN
TAJIKISTAN
GHANISTAN
PAKISTAN
TIBET
CHINA
Lhasa
Dharamsala
HIMALAYAN MOUNTAINS
NEPAL
BHUTAN
BANGLADESH
INDIA
MYANMAR
VIETNAM
LAOS
THAILAND
CAMBODIA
MALAYSIA
ARABIAN SEA
BAY OF BENGAL
INDIAN OCEAN

TUVA

Independent from 1921 until 1944, when it requested incorporation into the Soviet Union.

DECLARED:

14 August 1921

CAPITAL:	POPULATION:
Kyzyl	*310,000*
AREA:	CONTINENT:
170,500 km²	*Asia*
DISSOLVED:	LANGUAGE:
13 October 1944	*Tuvan, Russian*

Salchak Toka was a man with a mission. On graduation from Moscow's communist University of the Toilers of the East in 1929, he was keen to drag his homeland into the modern world. By following Soviet principles, his country – deep in the heart of Central Asia, sandwiched by the might of Russia and China – could make the transition from feudalism straight to socialism, bypassing capitalism along the way. Toka became General Secretary of the Central Committee of the Tuvan People's Revolutionary Party and set to work.

He made sure that Russians were granted full citizenship rights and traditional Tuvan religions systematically suppressed. Nomadic herding, the mainstay of Tuvan culture and livelihoods, was replaced by Soviet-style collective farms, while the Tibetan-Mongol script was discarded in favour of the Cyrillic alphabet. Toka won a Stalin Prize for a novel about herding, which some suggested he did not write, and received seven Orders of Lenin and the Order of the Red Banner of Labour. But Toka's most significant contribution was still to come.

It is 1944. Toka visits Moscow as the German army approaches the Soviet capital. He submits a petition requesting Tuva's admission to the Union of Soviet Socialist Republics. But war is being waged across the globe, and no one outside the Soviet world takes any notice. It is 1946 before they do, when Toka speaks on Soviet radio. He describes how his formerly independent country has been 'graciously accepted' by the Supreme Soviet. In the USSR, the press begins a campaign to popularize their new member. Rather than marking the second anniversary of Tuvan accession, it transpires that this is in fact the twenty-ninth anniversary: Russia's great October socialist revolution of 1917 is the key date that delivers on Tuva's 'century-old yearning for liberty and happiness'.

TUVA
Kyzyl
LAKE BAIKAL
RUSSIA
KAZAKHSTAN
Ulan Bator
MONGOLIA
Almaty
Urumqi
GOBI DESERT
KYRGYZSTAN
Beijing
CHINA
KISTAN
HIMALAYAN MOUNTAINS
NEPAL
New Delhi
BHUTAN
BANGLADESH
INDIA
MYANMAR
VIETNAM
LAOS
THAILAND
CAMBODIA

RYUKYU

Also known as Okinawa

Formerly independent kingdom seeking secession from Japan.

DECLARED:

4 February 2015

CAPITAL:

Naha

POPULATION:

1,400,000

AREA:

2,270 km²

CONTINENT:

Asia

LANGUAGE:

Japanese, Uchinaaguchi

Dr Yasukatsu Matsushima leaned across the conference table and with great ceremony passed a document to the man from the foreign ministry. In doing so, he made the first direct request to the Japanese government. As co-founder of an academic society for research into Ryukyuan independence, Dr Matsushima had just initiated what he intended to be an historic chain of events.

A major focus of Ryukyu's discontent revolves around the continued presence of US military bases on Okinawa, the main island in the Ryukyu archipelago. The bases are jointly administered by the Japanese and American governments, but neither side takes much notice of the local viewpoint. The once-independent Ryukyu islands were seized by Imperial Japan in the late nineteenth century and were devastated when US troops invaded in 1945, losing over a third of their population in the fighting. The Americans kept the islands afterwards, requisitioning land to build huge military facilities on Okinawa before finally returning the archipelago to Japan in 1972. But they hung on to the bases, which still occupy about one-fifth of the island.

Calls by Ryukyuan people to remove the Americans are routinely ignored by central government, but an independent Ryukyu could eliminate the military bases herself. Dr Matsushima and colleagues have built a strong legal case for a return to self-rule, based on careful research. The document passed to the foreign ministry on that damp afternoon in February 2015 protested that the forced annexation of Ryukyu was an obvious violation of international law and demanded that the colonization cease immediately.

The softly spoken professor of economics also asked for some documents in return: originals of three treaties between Ryukyu Kingdom and the USA, France and the Netherlands, signed with each country during the 1850s. These proved beyond doubt, he said, that Ryukyu was a sovereign nation at the time it was annexed.

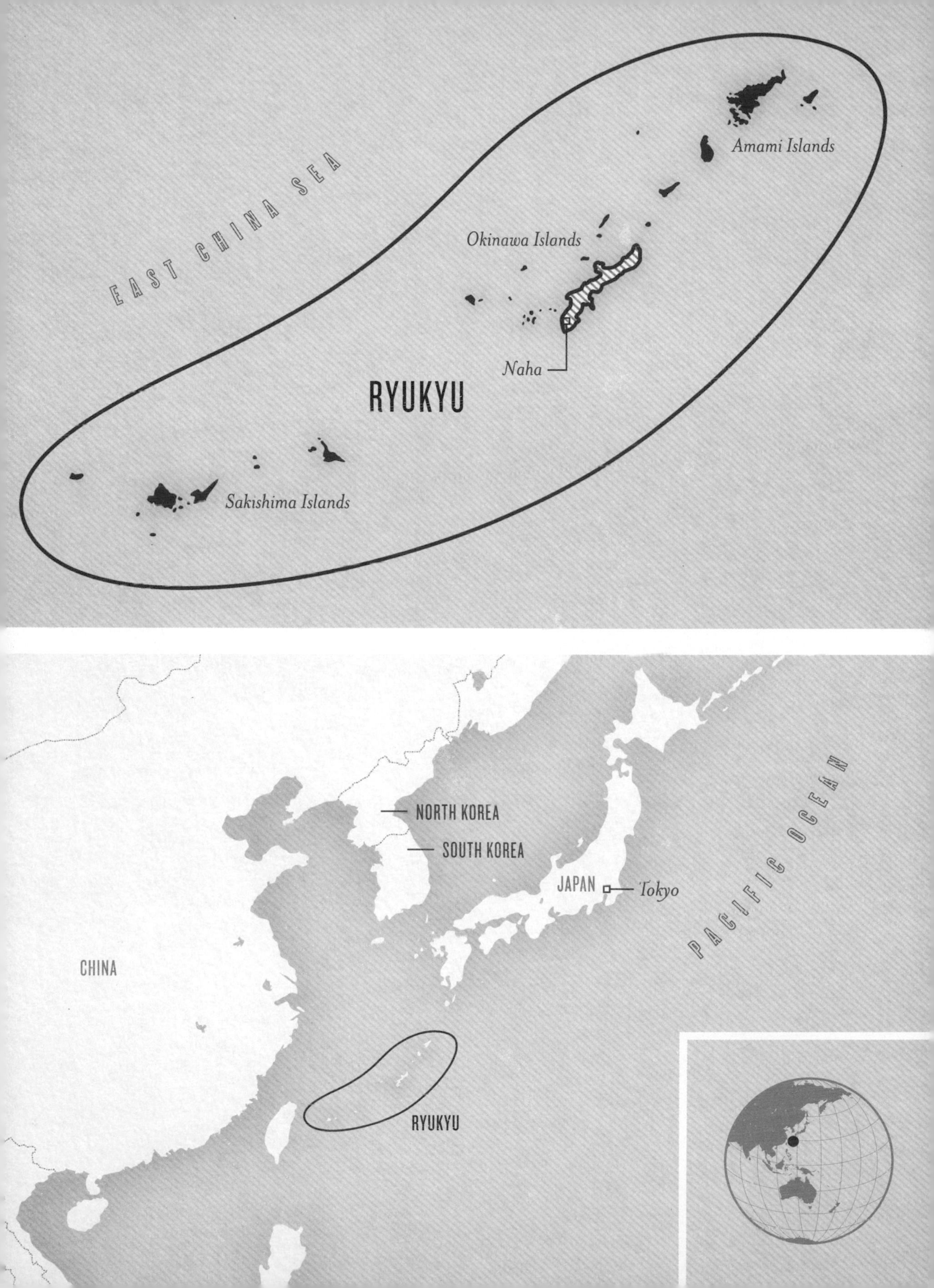

EAST CHINA SEA
Amami Islands
Okinawa Islands
Naha
RYUKYU
Sakishima Islands
NORTH KOREA
SOUTH KOREA
JAPAN
Tokyo
PACIFIC OCEAN
CHINA
RYUKYU

MOROC-SONGHRATI-MEADS

Also known as the Spratly Islands

Declared an independent kingdom in the 1870s by a British naval captain whose descendants have continued to dispute ownership of its islands with China, Taiwan, Vietnam, the Philippines and Malaysia.

DECLARED:

1877

CAPITAL:	POPULATION:
Southwark	*6,642*
AREA:	**CONTINENT:**
<50 km^2	*Asia*
FOUNDER:	**LANGUAGE:**
James George Meads	*English, Malay, Morac*

0 5 10 20 30 40 KILOMETRES

On Tuesday 5 July 1955, THE AGE, *an Australian newspaper, ran the following story.*

PLANES SEARCH FOR MYSTERIOUS ISLAND

MANILA, JULY 4 — THE PHILIPPINE AIR FORCE IS SEARCHING THE SOUTH CHINA SEA FOR A MYSTERIOUS ISLAND SETTLEMENT CALLED THE 'KINGDOM OF HUMANITY'.

THE PHILIPPINES PRESIDENT (MR RAMON MAGSAYSAY) SAID ON SUNDAY THAT HE WANTED TO KNOW WHETHER SUCH A PLACE ACTUALLY EXISTED. IF IT DOES, MR MAGSAYSAY WANTS TO DETERMINE WHETHER IT IS A LEGITIMATE SETTLEMENT WITHIN TERRITORIAL PHILIPPINES WATERS OR A CLANDESTINE BASE FOR SMUGGLERS AND COMMUNIST AGENTS.

SO FAR, THE ONLY FIRST-HAND DESCRIPTION OF THE 'KINGDOM' HAS COME FROM AN AMERICAN WHO CALLS HIMSELF 'CONSUL FOR THE SOVEREIGN'. MR MORTON F. MEADS, AGED 33, TOLD THE COMMANDER OF THE PHILIPPINES AIR FORCE, BRIGADIER-GENERAL PELAGIO CRUZ, THAT THE 'HUMANITY' SETTLEMENT HAS A POPULATION OF 3,400 INCLUDING CHINESE, AMERICANS, FRENCH, INDONESIANS, MALAYANS AND JAPANESE.

Meads traced his ancestry to James George Meads, master of the British ship MODESTE, *who in 1877 laid claim to an archipelago on behalf of the world's downtrodden and persecuted. Meads dedicated the island nation to a peaceful existence and proclaimed himself King James I.*

Morton Meads enlisted in the US army during World War II as the best way to fight for his country against the occupying Japanese force, and was cited for bravery in official dispatches. Since the war, the Philippines has not been the only country sniffing around Moroc-Songhrati-Meads, whose seabed contains rich deposits of oil and natural gas. In June 1972, Morton Meads led a delegation bound for the United Nations in New York to plead his nation's case against the region's bullies.

They never arrived. Tragedy struck when their ship, the E PLURIBUS UNUM, *was sunk by Category 1 Typhoon Ora off the Filipino coast. There was just one survivor: Morton Meads.*

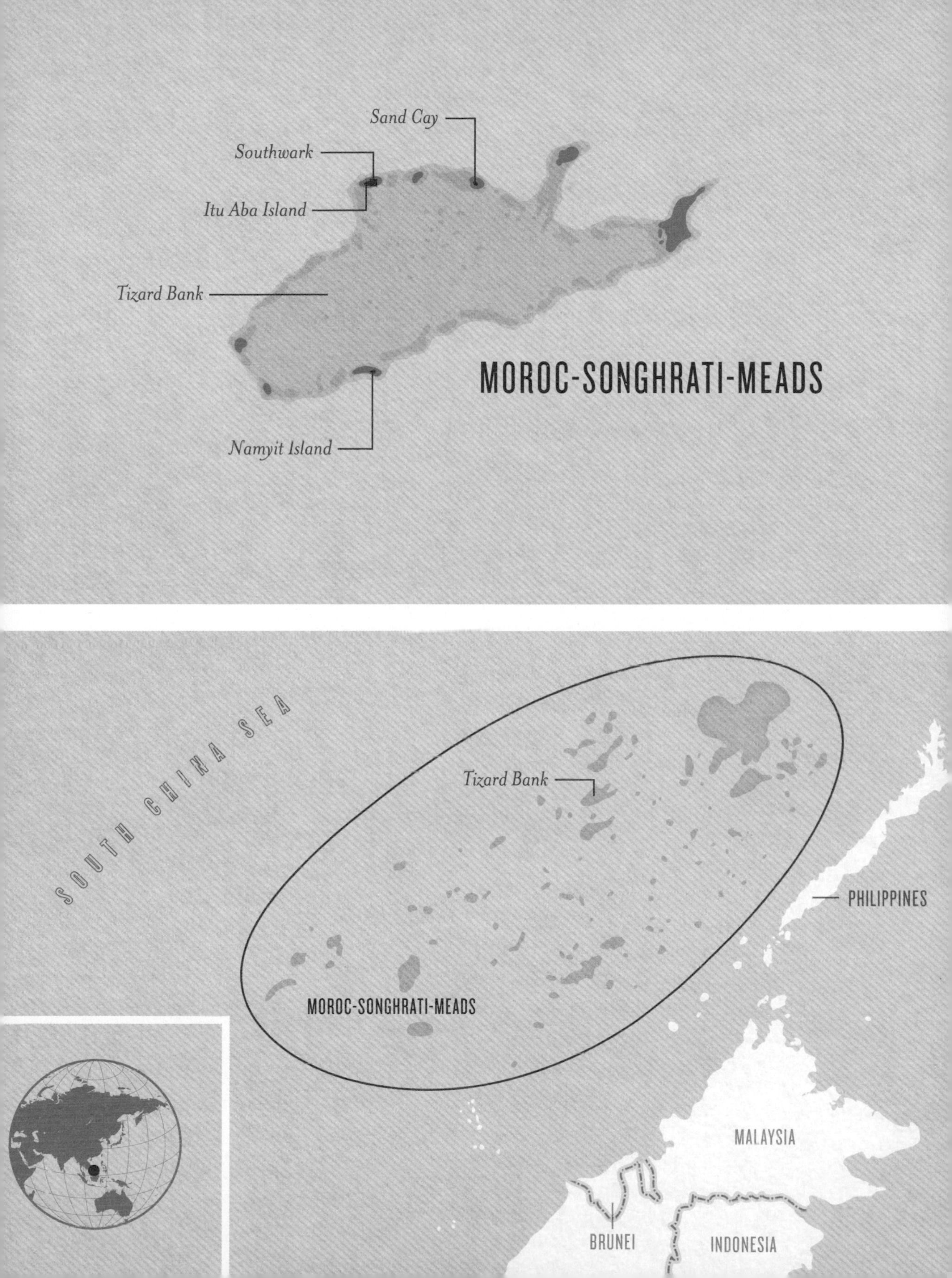

Sand Cay
Southwark
Itu Aba Island
Tizard Bank
MOROC-SONGHRATI-MEADS
Namyit Island
SOUTH CHINA SEA
Tizard Bank
PHILIPPINES
MOROC-SONGHRATI-MEADS
MALAYSIA
BRUNEI
INDONESIA

BANGSAMORO

Predominantly Muslim enclave in the southern Philippines.

DECLARED:

28 April 1974; 15 January 2012; 27 July 2013

CAPITAL:
Cotabato City

POPULATION:
3,900,000

AREA:
34,770 km²

CONTINENT:
Asia

LANGUAGE:
Moro-Maguindanoo, Filipino

The snipers were hidden in the coconut palms. They'd been waiting every night for a week, since the drones started flying over the village, keeping the children awake. Now they could see armed men approaching on foot across the maize fields.

By the end of the shooting, at least sixty-seven people were dead. Most were Filipino special police officers on a mission to arrest suspected terrorists early that Sunday morning. The police ran low on ammunition and called the army for reinforcements, which didn't arrive. The government called it a 'misencounter'. The media called it a massacre.

The village of Mamasapano is on the island of Mindanao, cornerstone of Moro Muslim territory. The Moros have a history of armed resistance to outside rule, beginning with the arrival of the Spanish in the sixteenth century. Many historians consider the Spanish-Moro War to be the world's longest anti-colonial struggle. It lasted 333 years. The fight against the Spanish was followed by the fight against the Americans and then, in turn, the Japanese. The war of independence against the Philippine state is just the latest phase of a 400-year-long national liberation movement for the Moro nation, or Bangsamoro. This is a Muslim–Christian conflict with very deep roots.

All of which was supposed to have been resolved in 2014 with the signing of an historic peace agreement. The creation of Bangsamoro, a new autonomous political entity, had been accepted in lieu of full independence. At last the poorest region in the Philippines had a chance to reverse its stunted growth. But the signal for an end to the unrelenting conflict had not been received by all. The Mamasapano massacre of January 2015 left forty-four police officers dead, the worst disaster for Filipino forces in decades. Another sad chapter in a war that was supposed to be over.

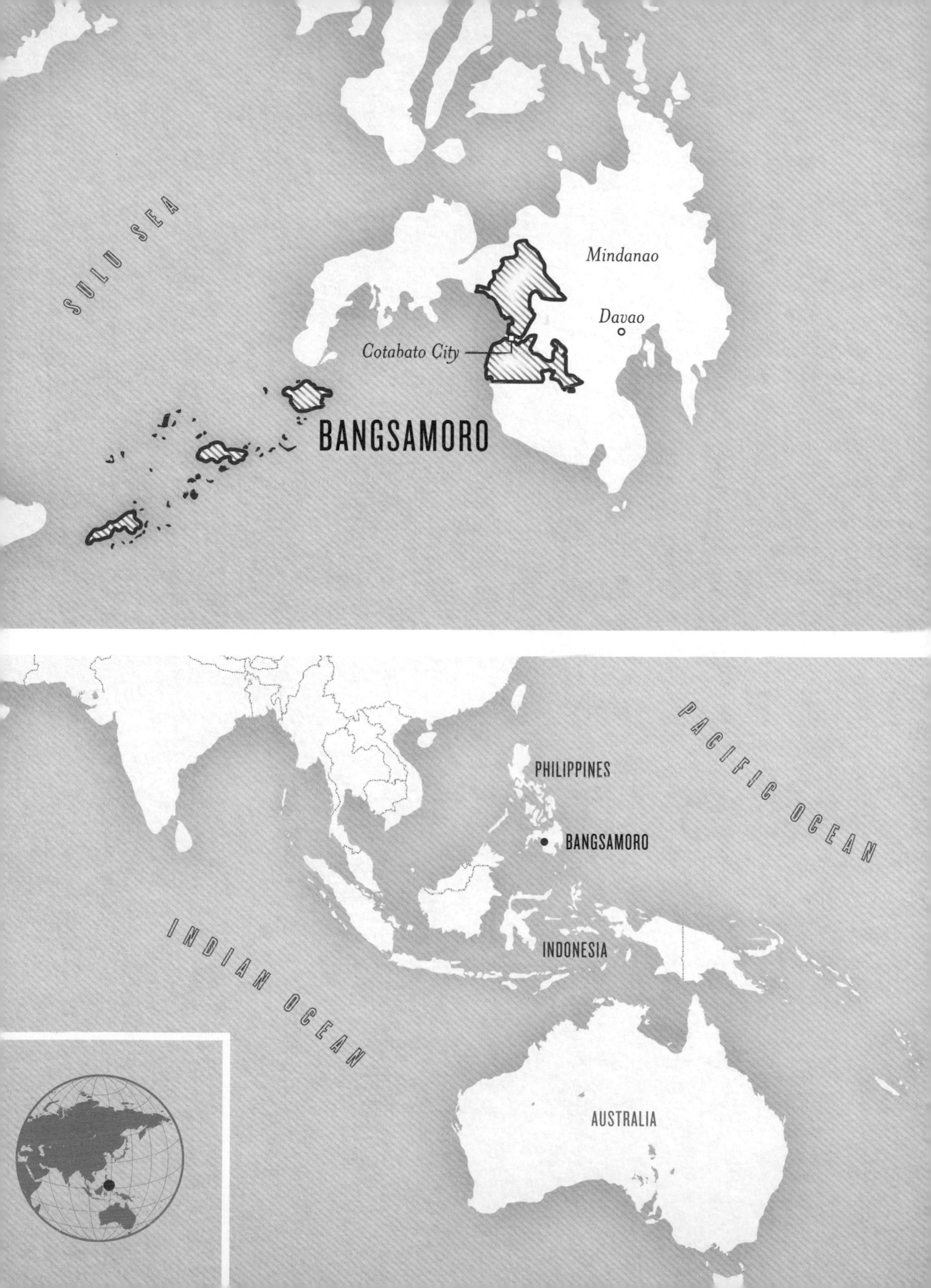

SULU SEA
Mindanao
Davao
Cotabato City
BANGSAMORO
PACIFIC OCEAN
PHILIPPINES
BANGSAMORO
INDIAN OCEAN
INDONESIA
AUSTRALIA

OCEA

ANIA

WEST PAPUA

Former Dutch territory annexed by Indonesia in 1963.

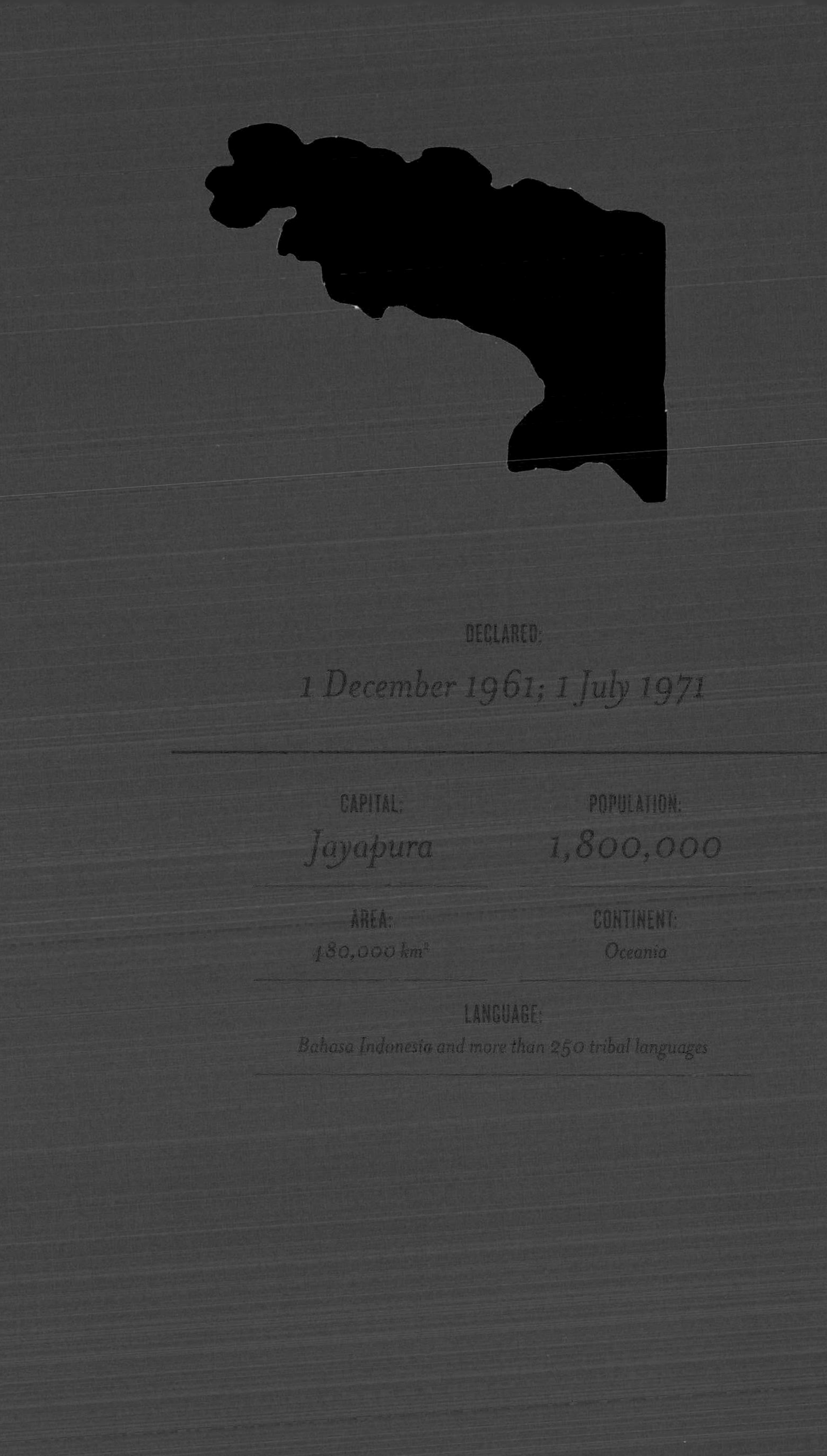

DECLARED:

1 December 1961; 1 July 1971

CAPITAL:

Jayapura

POPULATION:

1,800,000

AREA:

480,000 km²

CONTINENT:

Oceania

LANGUAGE:

Bahasa Indonesia and more than 250 tribal languages

Many believe that Theys Eluay was buried without his heart. It had been removed and sent away for analysis, they say, because the authorities thought he died of a heart attack. With or without it, he was buried at the local football ground. Twenty thousand people stood in the stifling heat to bid him farewell.

Mr Eluay was an unlikely independence hero, having once been a champion of integration with Indonesia. At the end of colonization in the early 1960s, the Dutch left and the Indonesians took over, promising West Papua a plebiscite on their future before the decade was out. When it came, the referendum took an unusual form. Eluay was one of just 1,025 tribal leaders hand-picked by the Indonesian government to represent nearly a million Papuans. To a man, the tribal leaders all voted to remain part of Indonesia. Years afterwards, Mr Eluay said the so-called Act of Free Choice had not felt particularly free.

Only much later did Eluay actively support independence. For a few short years his charismatic leadership appeared to unnerve the authorities. At the time of his death in 2001, he was free only on bail, charged with rebellion. The police cited a pro-independence gathering at his home and his role in a flag-raising ceremony. Flying the West Papuan 'Morning Star' flag in public still attracts a prison sentence of up to fifteen years.

They found his body the day after he attended a reception at the local headquarters of Kopassus, the Indonesian army's Special Forces. At 10 o'clock that night, Mr Eluay's driver made a distressed call to Eluay's home saying that his boss had been abducted. As the driver was speaking, the phone was cut off. The driver was never seen again. Seven Kopassus officers were convicted of Eluay's murder; the longest sentence was three years.

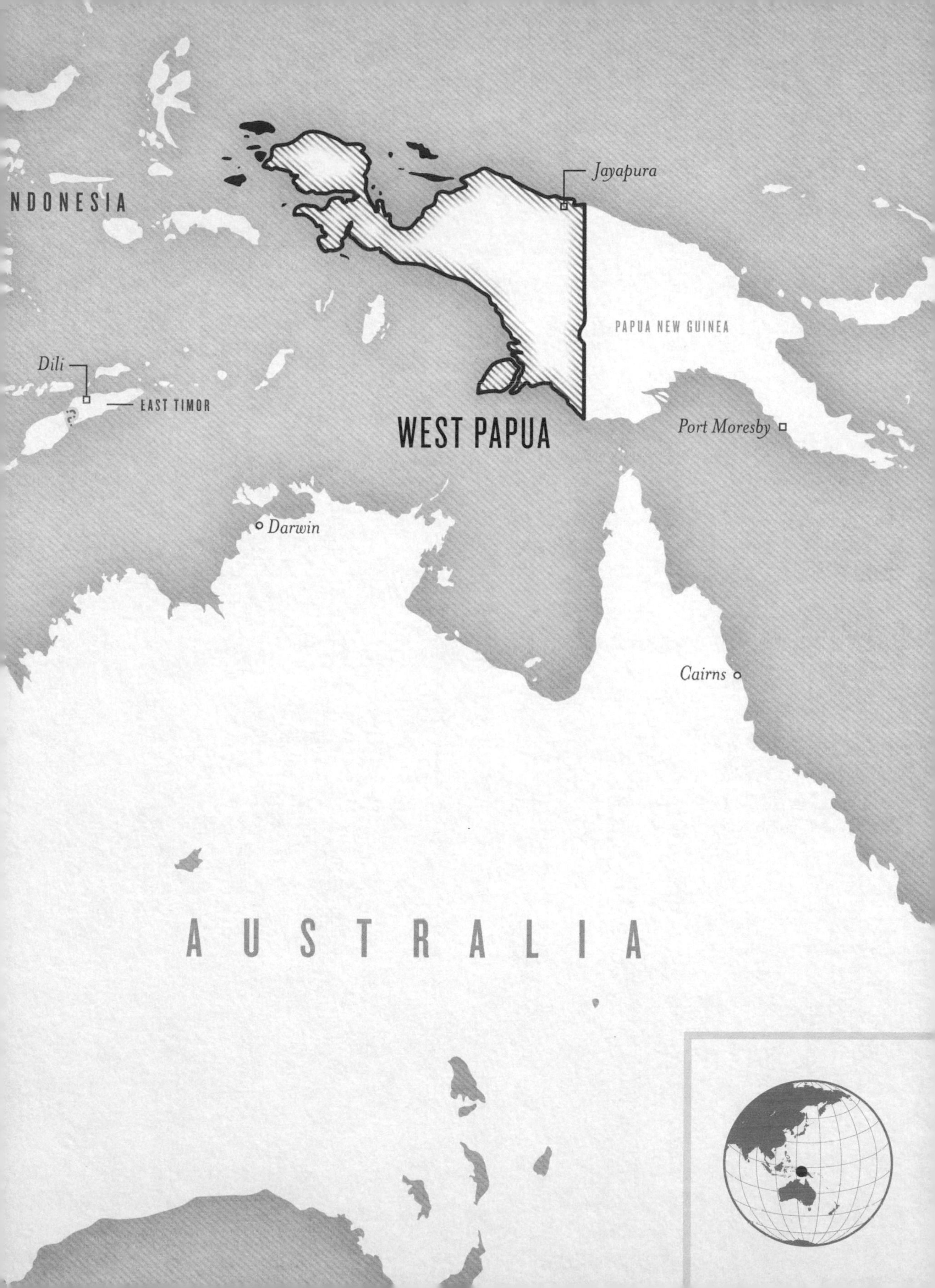
NDONESIA
Jayapura
PAPUA NEW GUINEA
Dili
EAST TIMOR
WEST PAPUA
Port Moresby
Darwin
Cairns
AUSTRALIA

MINERVA

Libertarian republic declared in 1972 on reclaimed land above a submerged Pacific atoll.

DECLARED:

19 January 1972

POPULATION:

0

AREA:

<1 km²

CONTINENT:

Oceania

DISSOLVED:

15 June 1972

LANGUAGE:

English

The old King of Tonga was no shrinking violet. At his peak, Taufa'ahau Tupou IV weighed in at 35 stone and was, according to the GUINNESS BOOK OF RECORDS, the world's heaviest sovereign: the largest king of the smallest kingdom. His Majesty was a jovial man, the benign ruler of a country known since the days of Captain Cook as the Friendly Islands.

But on 15 June 1972, His Majesty was feeling far from friendly. Months earlier, two nearby submerged reefs had been built up from concrete and coral blocks into very small islands by an international libertarian society, and declared the Republic of Minerva. The reefs had long served as fishing grounds for the Tongan people, and so it was that the king set off to reclaim his territory. He boarded the Tonga Shipping Company's vessel OLOVAHA, a former English ferryboat that used to ply the waters between Penzance and the Scilly Isles. With the royal standard flying from its mast, His Majesty filled the small boat with dignitaries in traditional tree-bark mats, worn around the waist, a platoon of soldiers in full dress uniform and the police brass band.

Two days later, at both South and North Minerva, the OLOVAHA's lifeboats were launched while the king remained aboard. But the final stretch before each island was too shallow, forcing the splendid royal entourage to wade ashore across the coral.

The Minervan flag was duly removed, hymns sung and prayers said. A royal proclamation was read as police and military stood, still dripping, to attention. A bugle sounded and, from across the water, the brass band played the national anthem. Tonga's red and white ensign was slowly raised and the soldiers fired a salute. The landfill project that had briefly become a republic was no more, now officially claimed part of the Kingdom of Tonga.

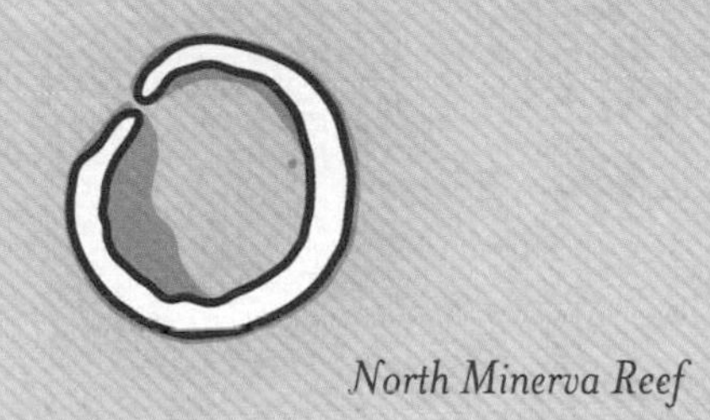

North Minerva Reef

MINERVA

SOUTH PACIFIC OCEAN

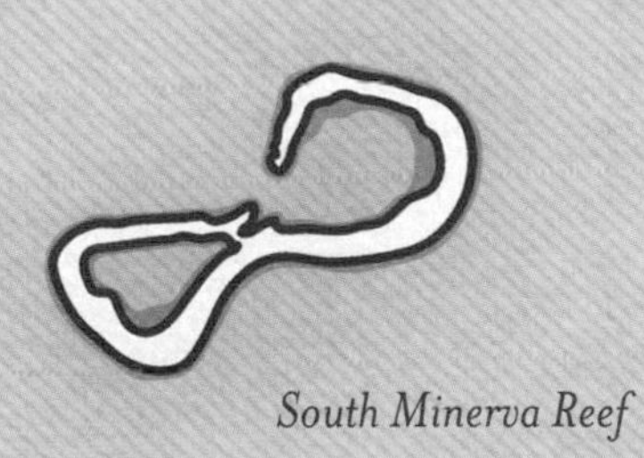

South Minerva Reef

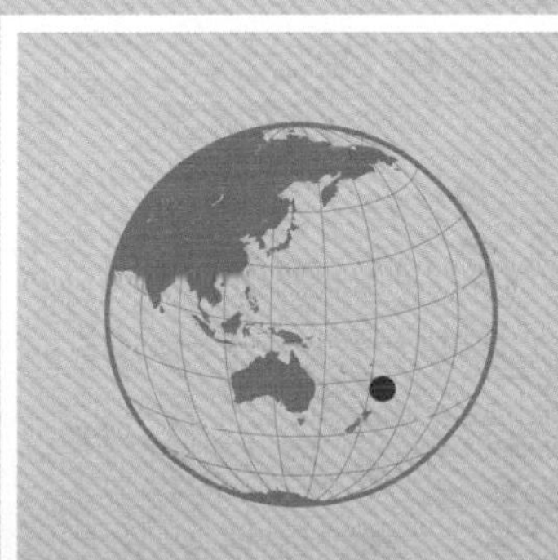

MURRAWARRI

Aboriginal people that never ceded sovereignty over their ancient homeland.

DECLARED:

30 March 2013

CAPITAL:	POPULATION:
Barringun	*1,500*
AREA:	**CONTINENT:**
81,796 km²	*Oceania*

LANGUAGE:

Muruwari, English

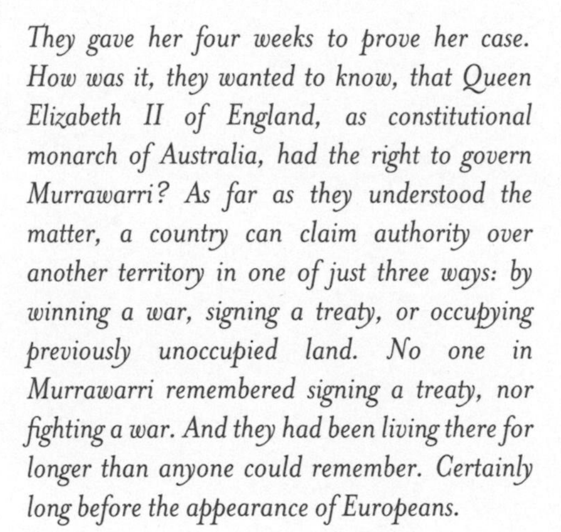

They gave her four weeks to prove her case. How was it, they wanted to know, that Queen Elizabeth II of England, as constitutional monarch of Australia, had the right to govern Murrawarri? As far as they understood the matter, a country can claim authority over another territory in one of just three ways: by winning a war, signing a treaty, or occupying previously unoccupied land. No one in Murrawarri remembered signing a treaty, nor fighting a war. And they had been living there for longer than anyone could remember. Certainly long before the appearance of Europeans.

Look at the fish traps down at Brewarrina. Nobody knows how many centuries they've been there, but they may be 40,000 years old: the most ancient human construction in the world. It is said those rocks were positioned by Baiame, god of the Muruwari tribe, creator of everything. He designed those traps for local people during a time of drought, by casting his great net across the creek. The rock walls and stone pens mimic the pattern of his net along 400 metres of the river bed. Murray cod, callop, catfish and silver perch could be herded and caught, during high and low flows alike, thanks to the stone traps. Steeped in legend and imbued with spiritual meaning, this was a bountiful spot in the dry scrub plain, one of the great inter-tribal meeting places of eastern Australia.

What then of that English monarch who would rule these prehistoric people? A month had passed and her time was up; answer came there none. For the people of Murrawarri, this resounding silence was tantamount to recognizing them as an independent nation. So in March 2013 the Murrawarri issued a declaration of the continuance of the state of their nation, a fundamental challenge to more than 200 years of colonial rule.

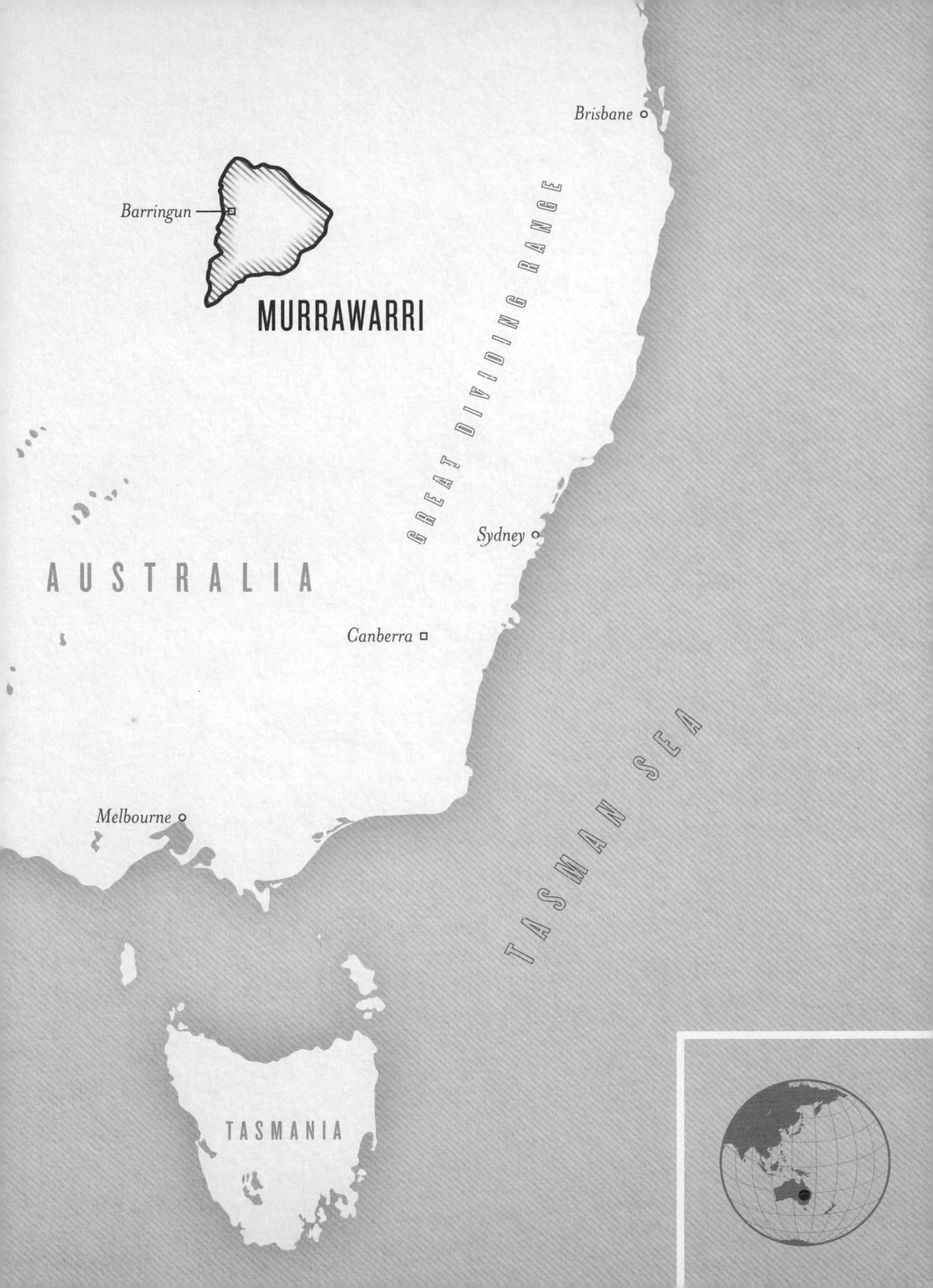

Brisbane
Barringun
MURRAWARRI
GREAT DIVIDING RANGE
Sydney
AUSTRALIA
Canberra
TASMAN SEA
Melbourne
TASMANIA

VEMERANA

Libertarian island republic briefly independent in 1980.

DECLARED:

28 May 1980

CAPITAL:

Luganville

POPULATION:

40,000

AREA:

3,956 km²

CONTINENT:

Oceania

DISSOLVED:

1 September 1980

LANGUAGE:

Bislama

FOUNDERS:

Michael Oliver and Jimmy Stevens

Four years in Nazi concentration camps had taught Michael Oliver all he needed to know about authoritarianism and the nation state. Although by now a naturalized US citizen, and a very wealthy one at that, he could still spot the signs of nascent repression. Capitalism in the USA wasn't quite as unfettered as he preferred. So he established the Phoenix Foundation to aid his pursuit of the ultimate libertarian state.

Enter Jimmy Stevens, one-time bulldozer operator, now full-time leader of a magico-religious cult. He had political aspirations for his homeland, the South Pacific island of Espiritu Santo, and they slotted neatly into Mr Oliver's agenda. Mr Stevens did a nice line in flamboyant oratory and liked to be known as Moses. In 1980, his island was part of an Anglo-French condominium heading for independence but Mr Stevens, with a little help from the Phoenix Foundation, had his own ideas.

At the head of a force armed solely with bows and arrows, Jimmy Stevens overcame the local administrators and declared the island a sovereign state. In line with years of colonial rivalry, the French recognized Vemerana, but the British did not. They prepared to send troops instead, until France objected. Meanwhile Mr Stevens concentrated on his libertarian principles. He had twenty-five wives.

But the rest of the Anglo-French archipelago soon became independent as the colonials had intended. Named Vanuatu, this new state was recognized by many more countries besides France. Vanuatu asked for assistance in the Mr Stevens affair from next-door Papua New Guinea, who readily agreed. When Papuan soldiers landed they confiscated wagonloads of bows and arrows and Vemerana's secession was soon over. Jimmy Stevens was arrested, tried and sentenced to fourteen years in prison. With his downfall came the end of Mr Oliver's attempts at nation-building. His dream of creating an autonomous laissez-faire utopia reverted to just that.

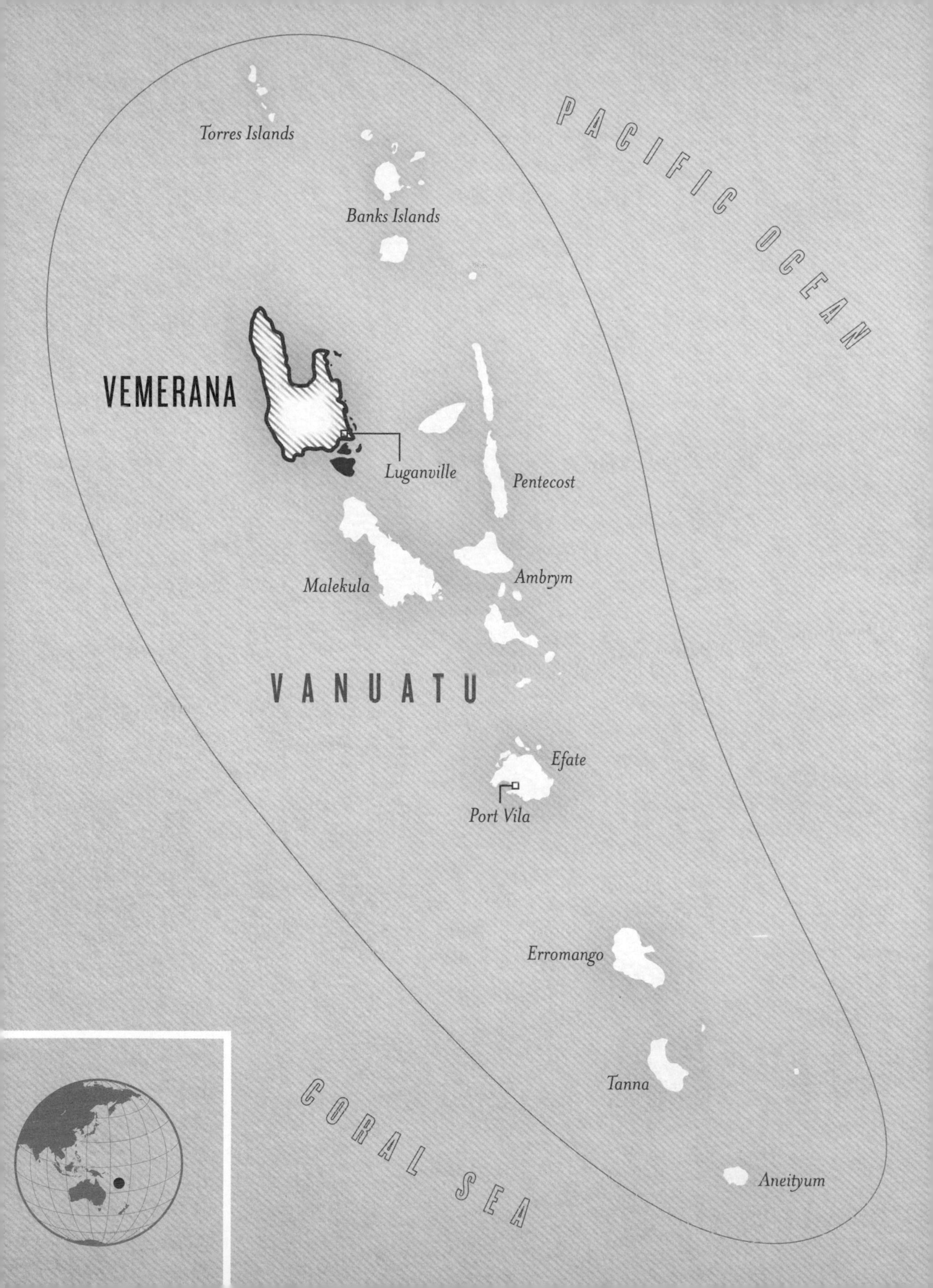
PACIFIC OCEAN
Torres Islands
Banks Islands
VEMERANA
Luganville
Pentecost
Malekula
Ambrym
VANUATU
Efate
Port Vila
Erromango
Tanna
CORAL SEA
Aneityum

HUTT RIVER

Seceded from Australia in 1970 in protest against newly introduced grain quotas.

DECLARED:

21 April 1970

CAPITAL:	POPULATION:
Nain	*23*
AREA:	**CONTINENT:**
75 km²	*Oceania*
FOUNDER:	**LANGUAGE:**
Leonard Casley	*English*

On 27 May 1976, a cablegram was sent by the Australian department of foreign affairs in Canberra to the prime minister. Marked 'Restricted', its subject was Hutt River Province:

OUR LEGAL ADVICE IS THAT THE INFORMATION AND EVIDENCE AVAILABLE TO DATE DOES NOT SUGGEST THAT CASLEY HAS CONTRAVENED ANY AUSTRALIAN LAW. HOWEVER, CASLEY'S ACTIVITIES TO OBTAIN RECOGNITION FOR HIMSELF AS 'PRINCE LEONARD' AND FOR 'HUTT RIVER PROVINCE' AS SOVEREIGN HAVE BEEN EXAMINED CAREFULLY AND, WHERE APPROPRIATE, ADMINISTRATIVE ACTION TAKEN TO COUNTER THEM BY THE COMMONWEALTH AND WESTERN AUSTRALIAN GOVERNMENTS.

Picture gentle, rolling terrain just 30 kilometres inland from the Indian Ocean, on the western edge of the huge Australian landmass. Fields of snow-white and sky-blue lupins, grown as a high-protein food for livestock, stretch to the horizon. The year is 1970 and farmer Leonard George Casley, his wife Shirley and seven children are alone in a kingdom of their own making. After many attempts to reverse a strict new quota on their production of wheat, Casley has taken the ultimate decision: to split from Australia altogether.

Over the years, Hutt River Principality has had many spats with its larger neighbour. The Australian postal service refuses to handle outgoing Hutt River mail, forcing it to be diverted via Canada. Following repeated tax demands, Mr Casley officially declares war on Australia but is ignored. He declares a cessation of hostilities within the week. Hutt River is not recognized but gains an acceptance of sorts. In their Notices of Assessment from the Australian government's taxation office for 2005, Leonard and Shirley are deemed to be non-residents of Australia. Neither has anything to pay. A reproduction of Prince Leonard's tax return is mounted on the wall outside the Nain post office for all to see. It is one of Hutt River's prized landmarks.

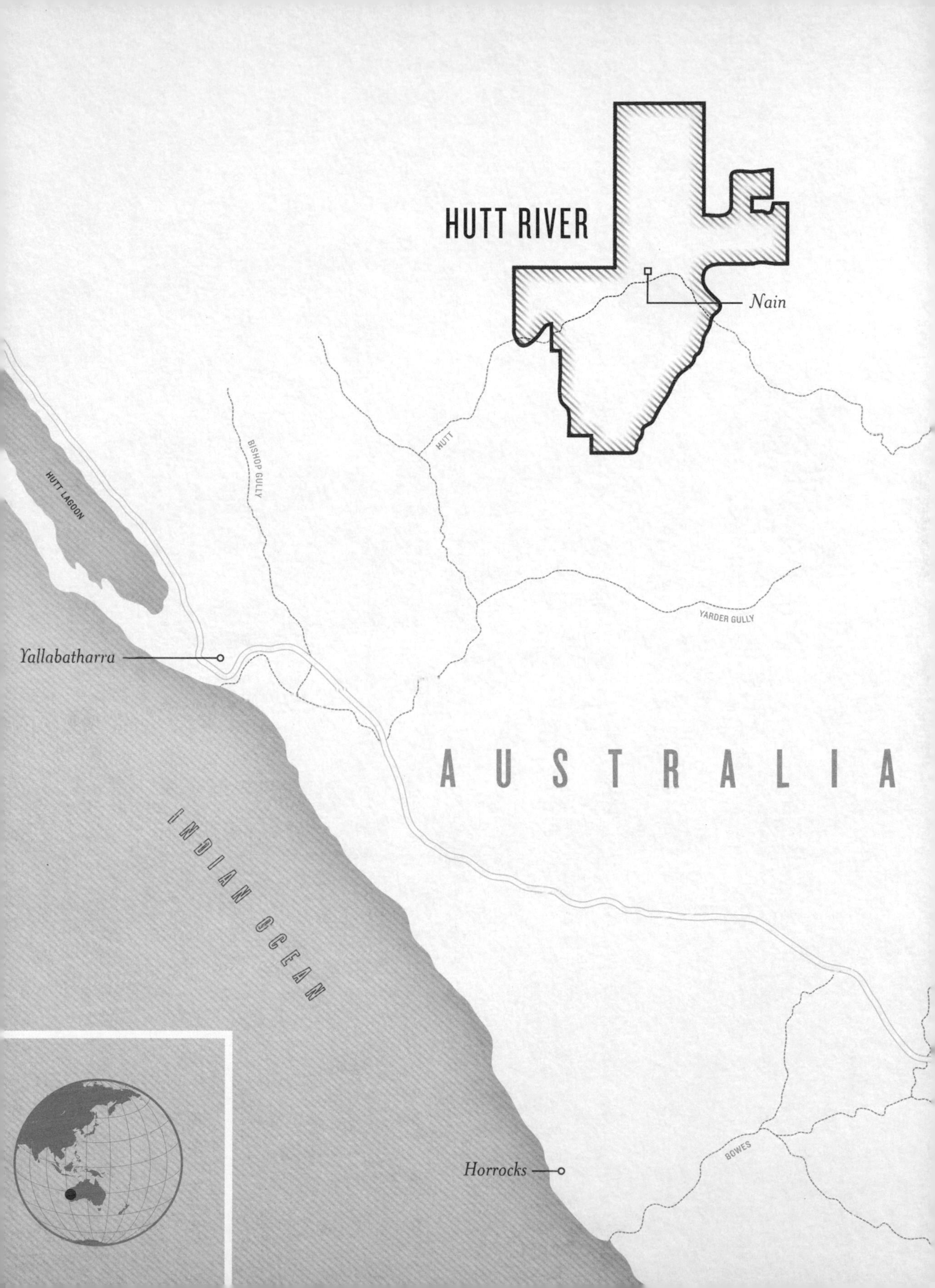
HUTT RIVER
Nain
HUTT
BISHOP GULLY
HUTT LAGOON
YARDER GULLY
Yallabatharra
AUSTRALIA
INDIAN OCEAN
BOWES
Horrocks

ELSEW

HERE

ATLANTIUM

A primarily non-territorial state – albeit based in Australia – founded in 1981.

DECLARED:

27 November 1981

CAPITAL:

Concordia

POPULATION:

>2,000

citizens in over 100 countries

AREA:

0.76 km²

CONTINENT:

Non-territorial

FOUNDERS:

George Cruickshank, Geoffrey Duggan and Claire Duggan

LANGUAGE:

English, Latin

This is a state, but not as you know it. In a leafy suburb of Sydney, Australia, a self-declared sovereign entity sets out to challenge conventional notions of what makes a country a country. The idea that a nation-state be determined by its geographical boundaries is running out of steam. In the Empire of Atlantium they believe that territorial claims as the basis for legitimate statehood are all but meaningless in the modern globalized world.

A state is made up of people, and people move – today more than ever before. A community of citizens – wherever they happen to be distributed geographically – forms the primary legitimacy of Atlantium. Its sovereignty is founded upon the will of its people. This is an extraterritorial, transnational, intercultural state.

Within months of its proclamation in November 1981, the original inhabitants of Atlantium had drawn up a constitution. They decreed a new calendar, which begins at the end of the last Ice Age. Latin became one of its official languages because today it is a tongue that has little association with any living culture. Stamps, coins and banknotes also appeared, the currency featuring images of the Imperial Eagle and Atlantium's head of state, Emperor George II (born George Cruickshank in Sydney, Australia).

From its beginnings in a Sydney apartment, Atlantium moved to occupy a rural province with a fenced perimeter. This is the empire's global administrative capital, ceremonial focal point and spiritual homeland. A territory for a non-territorial state. Like Rome, the province occupies seven hills. The rolling countryside is occupied by more kangaroos than humans.

Unlike most other countries, the Empire of Atlantium does not issue passports. It advocates the unrestricted international freedom of movement for all peoples and therefore neither issues nor recognizes any form of restrictive travel documentation. Citizens of Atlantium are true citizens of the world.

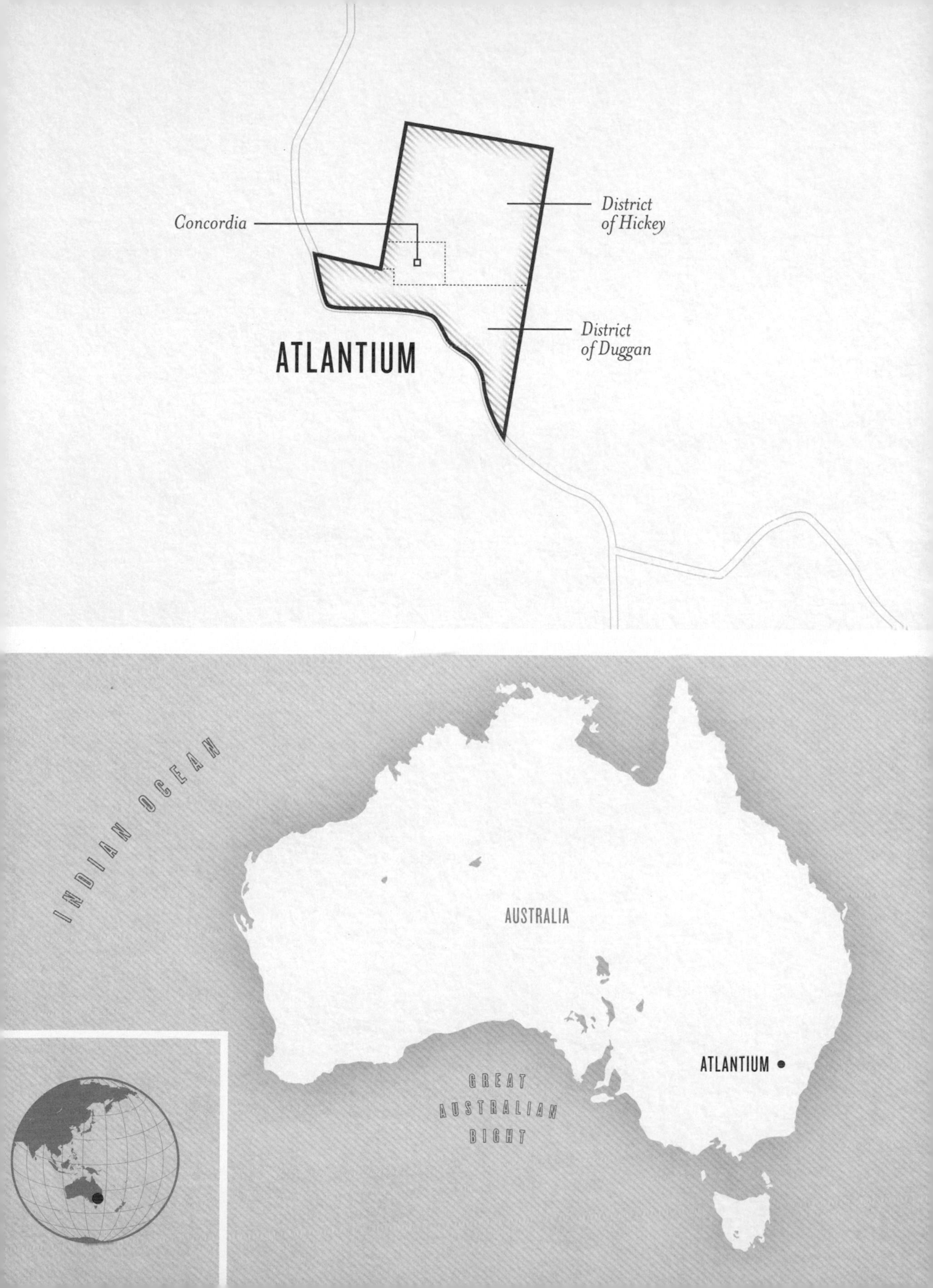
Concordia
District of Hickey
District of Duggan
ATLANTIUM
INDIAN OCEAN
AUSTRALIA
ATLANTIUM
GREAT AUSTRALIAN BIGHT

ANTARCTICA

The 'last continent' where disputes over territorial sovereignty are uniquely set aside.

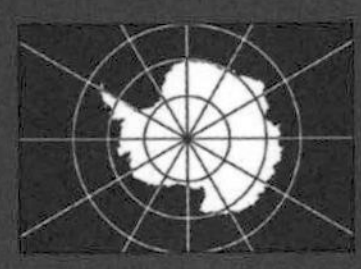

DECLARED:

1 December 1959

POPULATION:

0

(5,000 temporary residents)

AREA:	CONTINENT:
14,000,000 km²	*Antarctica*

This is a land of extremes, harsh, remote and unforgiving. A land of active volcanoes and spectacular mountain ranges but almost entirely wreathed in ice. The coldest, windiest and driest continent; one with no trees or bushes, just lichen and moss. No indigenous population; no permanent human population at all. The first woman to arrive was in 1935; the first child born not until 1979. A frozen wilderness, larger than Europe, inhabited mainly by penguins.

All of which helps to explain why it also differs politically. Under normal circumstances, this territory would have been carved up long ago. It began like that, in the early twentieth century, when imperialism was still a rampant force. They helped themselves to slices, like a frozen dessert. Until they stopped. Some wedges of the Antarctic cake remained unclaimed by any country, but several states had made claims and some of those claims overlapped. By the middle of the twentieth century, these territorial positions had been asserted, but not agreed. Tensions arose, until cooperation prevailed, and an international treaty was signed in 1959.

Some countries continue to explicitly recognize the territorial claims, some maintain a policy of not recognizing any claims, and some reserve the right to make their own claim. Meanwhile, the goal of sovereignty is on hold. The Antarctic transcends the norms of the nation state, an ice-bound challenge to the global standard for territorial control. This is the exception that proves the rule. The entire continent is used exclusively for peaceful purposes, the preserve solely of scientists and adventure tourists. All thanks to the Antarctic Treaty, a unique agreement to govern a unique place.

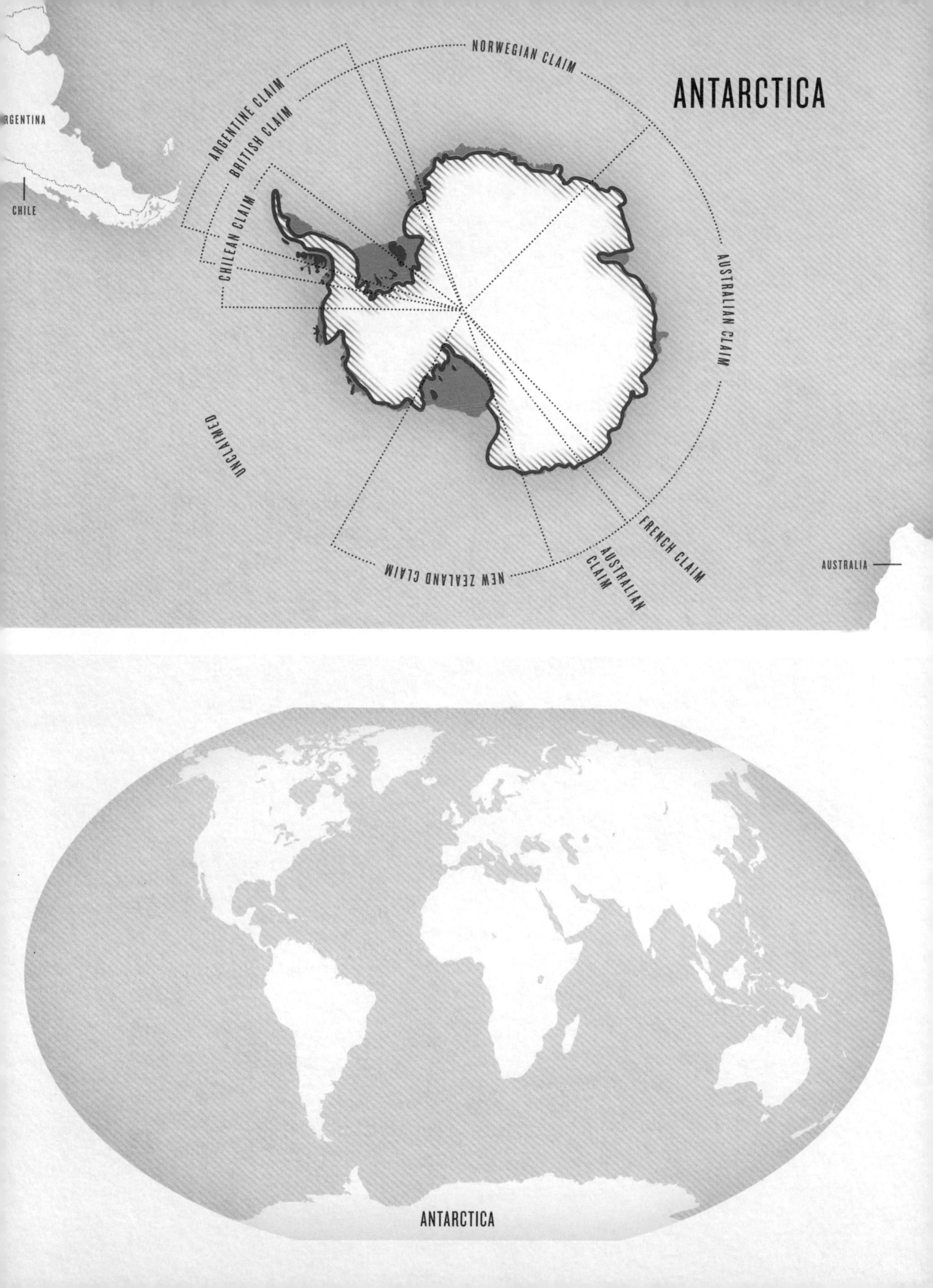
ANTARCTICA
NORWEGIAN CLAIM
ARGENTINE CLAIM
BRITISH CLAIM
CHILEAN CLAIM
AUSTRALIAN CLAIM
FRENCH CLAIM
AUSTRALIAN CLAIM
NEW ZEALAND CLAIM
UNCLAIMED
RGENTINA
CHILE
AUSTRALIA
ANTARCTICA

UNITED MICRONATIONS MULTI-OCEANIC ARCHIPELAGO (UMMOA)

A multi-oceanic archipelago – twenty-nine insular possessions and one Antarctic continental territory – leading to territorial disputes with Bangladesh, Belize, Colombia, Comoros, France, Haiti, India, Jamaica, Kiribati, Madagascar, Marshall Islands, Mauritius, Nicaragua, Portugal, Seychelles and the USA.

DECLARED:

6 May 2008

CAPITAL:

Cyberterra

POPULATION:

67

AREA:

432,000 km²

CONTINENT:

Multi-continental

LANGUAGE:

English, Italian

0 0.5 1 2 3 KILOMETRES

The initial island members of UMMOA were grouped together for statistical convenience only, assembled by the International Organization for Standardization for bureaucratic reasons. A ragtag collection of uninhabited atolls, islets and reefs spread across a vast area of the Pacific Ocean, they had little in common other than being under the sovereignty of the USA. This union was officially assigned a two-letter code, or alpha-2 as they call it: UM. A disparate quirk of political geography, they were otherwise known as the United States Minor Outlying Islands.

These isles were promptly invaded in a virtual fashion, collectively given an internet country code, '.um', in line with their alpha-2; but this internet country code was abandoned because the domain was not used. Like a seaworthy ship discarded in international waters, the abandoned code was legally occupied in May 2008 and the United States Minor Outlying Islands annexed. They were renamed the United Micronations Multi-Oceanic Archipelago. UMMOA for short. UMMOA has grown and diversified. Its capital, the city of Cyberterra, was founded as a metrosite or virtual realm, given terrestrial coordinates of 43° 0' 0" North, 15° 0' 0" East, a point in the Adriatic Sea. The multi-oceanic archipelago went continental, laying claim to a slice of Antarctica. It asserted its mandate over a piece of the Great Pacific Garbage Patch and several islets threatened by rising sea-levels in the Pacific and Indian Oceans.

As some territory disappears, elsewhere new land emerges. Deep beneath the gentle waves of the Red Sea, tectonic rumblings began late in 2011. Fishermen from the port of Salif in Yemen spied a fountain of blood-red lava spurting 30 metres into the air, and a new island was duly formed in the Zubair archipelago. UMMOA named it and claimed it; Aphrodite Island became UMMOA's most recent national component.

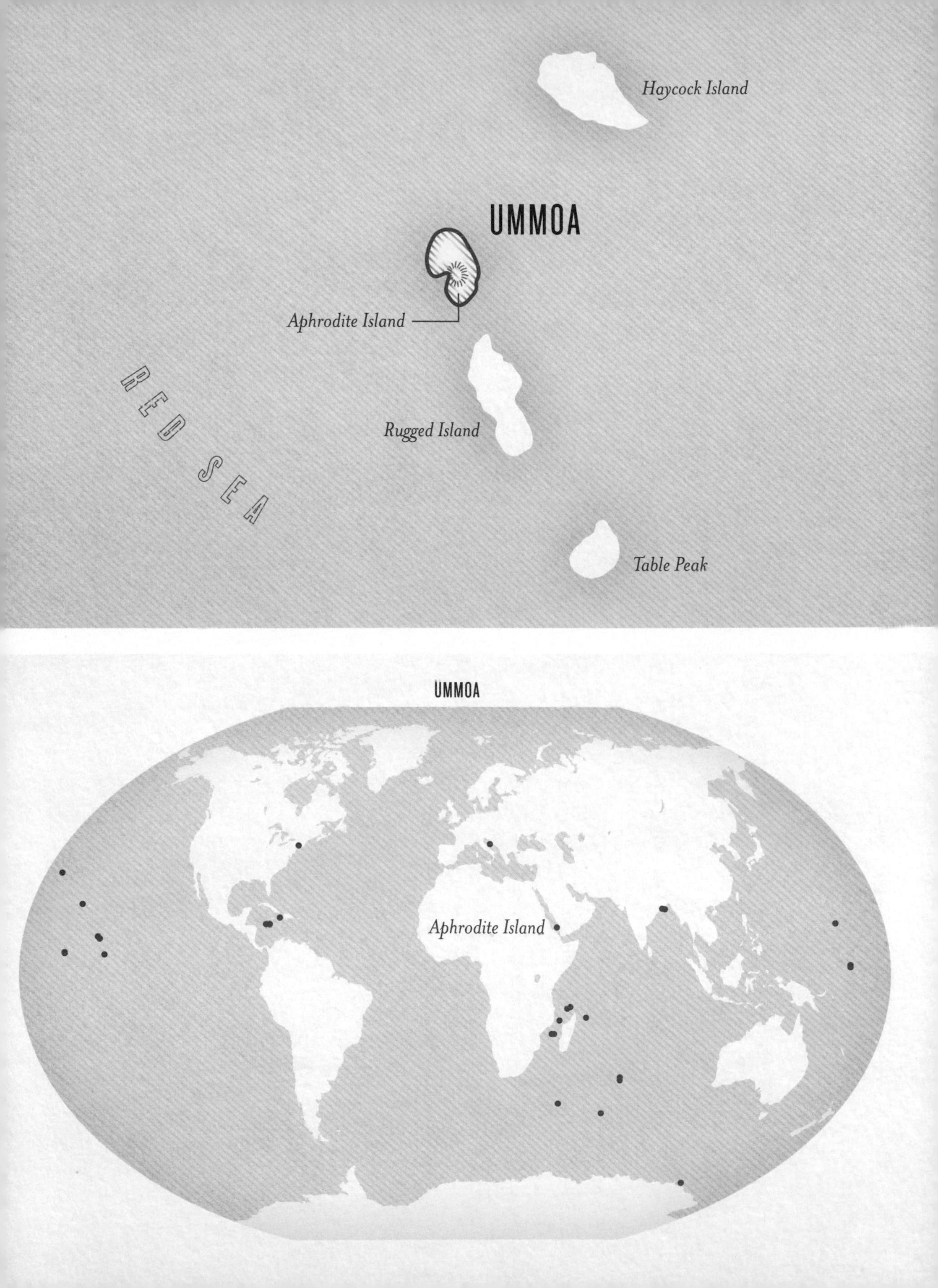
Haycock Island
UMMOA
Aphrodite Island
RED SEA
Rugged Island
Table Peak
UMMOA
Aphrodite Island

ELGALAND-VARGALAND

Proclaimed in 1992, incorporating all boundaries between other nations as well as digital territory and other states of existence, such as the dream state.

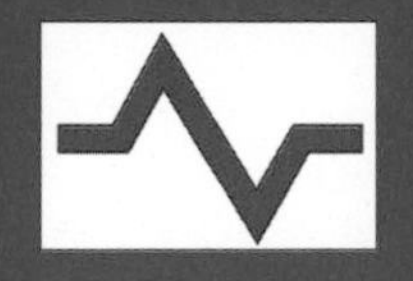

DECLARED:

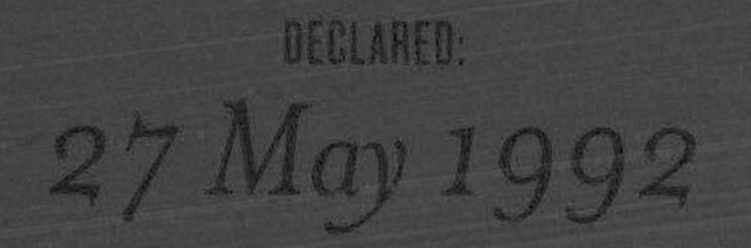

27 May 1992

POPULATION:

Most populous realm on Earth

AREA:

All border territories – geographical, mental and digital

FOUNDERS:

Carl Michael von Hausswolff and Leif Elggren

CONTINENT:

Multi-continental

LANGUAGE:

English, Swedish

One day two artists, Carl Michael von Hausswolff and Leif Elggren, met in Stockholm, Sweden and agreed to establish a new country. They had no army to back up their territorial ambitions, but they had identified spaces where they could establish their state without ruffling too many feathers.

They began with fragments of land. Like the forgotten scraps of food left on your plate after a meal, these slivers left over between countries can be called No Man's Land no longer: Von Hausswolff and Elggren have annexed them as part of Elgaland-Vargaland.

Like any empire, this country is in flux. New territories appear overnight, as along the border between Sudan and South Sudan, annexed as soon as South Sudan declared independence in 2011. Elsewhere, such unclaimed terrain disappears as when the border between East and West Germany dissolved. Visitors to Elgaland-Vargaland are numbered in their millions every year. Each time you travel somewhere you visit Elgaland-Vargaland.

On the kingdom's tenth anniversary in 2002, a group of citizens boarded the ferry from Stockholm to Tallinn, Estonia. They carried only their Elgaland-Vargaland passports. The Estonian authorities held them for a day for trying to enter the country with false documents before they were sent back to Sweden. Their purpose was to go home to Elgaland-Vargaland, to be in turn rejected at the Swedish border, then returned to Estonia and so on, for ever. Their passports were confiscated before their return to Sweden, so they visited home for only one day.

However, the country comprises more than simply physical space, incorporating digital territory and other states of existence entirely. When the Roman Catholic Church effectively buried the concept of limbo – the place where babies who died without baptism went – it too was annexed. Each time you enter another form, such as the daydream state, you become another visitor to Elgaland-Vargaland.

ACKNOWLEDGEMENTS

I'd like to thank a number of colleagues at Oxford University's School of Geography and the Environment whose assistance with, and thoughts on, this project are much appreciated. They are Simon Abele, Henri Rueff, Tim Schwanen and Fiona McConnell. My agent, Gordon Wise, worked hard to see this book into fruition, as did Jon Butler, publisher and editor, who had faith and enthusiasm as well as an eagle eye. I am particularly grateful to Sarah Greeno for her hard work and attention to detail in producing the maps and designing the book. Finally, my thanks also go to Lorraine Desai and Mia Middleton for tolerating the long evenings I was locked away in my office.